Overcurrent Protection
NEC Article 240 and Beyond

RIVER PUBLISHERS SERIES IN POWER

The "River Publishers Series in Power" is a series of comprehensive academic and professional books focussing on the theory and applications behind power generation and distribution. The series features content on energy engineering, systems and development of electrical power, looking specifically at current technology and applications.

The series serves to be a reference for academics, researchers, managers, engineers, and other professionals in related matters with power generation and distribution.

Topics covered in the series include, but are not limited to:

- Power generation;
- Energy services;
- Electrical power systems;
- Photovoltaics;
- Power distribution systems;
- Energy distribution engineering;
- Smart grid;
- Transmission line development.

For a list of other books in this series, visit www.riverpublishers.com

Overcurrent Protection NEC Article 240 and Beyond

Gregory P. Bierals
Electrical Design Institute, USA

Published 2022 by River Publishers
River Publishers
Alsbjergvej 10, 9260 Gistrup, Denmark
www.riverpublishers.com

Distributed exclusively by Routledge
605 Third Avenue, New York, NY 10017, USA
4 Park Square, Milton Park, Abingdon, Oxon OX14 4RN

Library of Congress Cataloging-in-Publication Data

Overcurrent Protection NEC Article 240 And Beyond / Gregory P. Bierals.

Routledge is an imprint of the Taylor & Francis Group, an informa business

ISBN 978-87-7022-621-9 (print)
ISBN 978-87-7022-620-2 (online)
ISBN 978-1-003-20729-0 (ebook master)

Contents

Preface

The earliest form of a fuse was developed as a means of protecting telegraph lines from the effects of lightning. The concept of using a fusible link, where electrical resistance is low enough to allow this element to act as a conductor, and yet, when current reaches destructive levels, this link will fuse or melt at a lower temperature than the protected conductor to open the circuit, has been with us for a very long time. In its original form, the fuse was nothing more than a wire with a smaller cross-sectional area than the protected conductor, thus allowing the smaller wire to fuse or melt and provide the desired protection.

The earliest form of a circuit breaker was developed by Thomas Edison. However, the introduction of this protective device did not come about until much later.

Indeed, as I was entering the electrical trade (August 3, 1964), and eager to learn about the electrical industry, I read an article in a monthly business publication which stated that if a circuit breaker interrupted a short-circuit, the tripping mechanism may have been damaged in the process, and it should be replaced. Another recommendation stated that circuit breakers installed in an industrial environment should be replaced at 5-year intervals as a routine maintenance procedure.

Now, I am not lending any credence to these ideas of that time. And this may have been a good idea. Certainly, the opening of a circuit breaker under short-circuit conditions will produce severe magnetic stresses on its internal components in accordance with the square of the instantaneous peak currents associated with the fault. And the heat energy will be relative to the square of the symmetrical current (rms) of the short-circuit. These conditions may be, and often are, damaging to this device. And in certain industrial environments, other conditions, including moisture, excessive heat, chemicals, and other factors may lead to damage.

And conditions have improved over the past 50 years. I have reviewed the testing procedure for Molded-Case Circuit Breakers (UL 489). And these overcurrent devices are truly remarkable. With interrupting ratings as high as

200,000 amperes and fault-clearing times of ½ cycle (.008 seconds), or less for current-limiting type circuit breakers.

And the testing standard of UL 198 for fuses provides guidance in proper application of these protective devices. Considering interrupting ratings as high as 300,000 amperes and fault-clearing times as low as 1/4 cycle (.004 seconds), these devices offer a level of protection unparalleled in the industry. And, considering the fact that there is no added impedance in the testing procedure assures that the identified interrupting rating is exactly the amount of current that the fuse was subjected to during the test.

Insulated-Case Circuit Breakers have a molded-case and a two-step stored-energy mechanism. In larger sizes, the instantaneous-trip setting can be adjusted. This also applies to Low-Voltage Power Circuit Breakers.

The application of fuses has evolved into an extremely wide range of protection standards over the course of many years. From the early days of providing protection against overloads by causing a fusible link to melt over time, to the development of current-limiting overcurrent devices that have interrupting ratings as high as 300,000 amperes. And operating characteristics that are relative to fault-clearing times of less than ½ cycle (short-circuit currents reach a peak during the first ½ cycle).

Selective coordination has become an important issue over the past several years. Multiple elevator cars supplied by a single feeder circuit, Information Technology Equipment, and Emergency Systems are examples where the localization of an overcurrent to restrict outages is mandatory. Published data from equipment manufacturers, such as selectivity ratios of line-side to load-side overcurrent device ratings help to simplify this form of protection.

And, finally, the concept of providing overcurrent protection in accordance with the short-circuit withstand rating of conductor insulation and equipment short-circuit current ratings has expanded in recent years.

Finally, assuring the adequacy of the effective ground-fault current path and its relationship to the selection of an overcurrent device is critical to the overall protection scheme.

These are examples of the subject matter in this book. Overcurrent protection may seem to be an easy subject. And, on the surface, I would certainly agree. But, delving further into this topic, I actually found how little I knew, and really, just how much that I had forgotten.

We begin with a list of terms and definitions (NEC Article 100), along with a detailed explanation of these defined terms. It is critical that the reader carefully review this information and any appropriate related information before continuing to the later chapters.

Throughout this book there are appropriate references to the National Electrical Code, most especially, Article 240. It would be a good idea to review these references which will help to reinforce the subject matter.

At the end, there is a list of 50 multiple choice questions relating to the NEC and an answer key to check your results.

Writing this book has been a truly enlightening and rewarding experience.

I humbly present this material, and, as always, I welcome your comments and suggestions.

Sincerely,
Gregory P. Bierals
Electrical Design Institute

1

Definitions

ACCESSIBLE, READILY (READILY ACCESSIBLE) - Capable of being reached quickly for operation, renewal, or inspections without requiring those to whom ready access is requisite to actions such as to use tools, to climb over or remove obstacles, or to resort to portable ladders, and so forth. Overcurrent devices are normally required to be readily accessible, and Section 240.24(A) states this requirement. Certain exceptions apply however, such as an overcurrent device installed on a busway (368.17(C)), feeder and service overcurrent devices, such as cable limiters which are installed at higher elevations. Or, where the feeder or service overcurrent devices may be locked or sealed and not readily accessible to a building occupant (225.40, 230.92), or for supplementary overcurrent devices, such as a supplementary overcurrent device in a lighting fixture (luminaire) (240.10), or where overcurrent devices are installed adjacent to equipment which is located at higher elevations and accessible by portable means.

AMPACITY - The maximum current, in amperes, that a conductor can carry continuously under the conditions of use without exceeding its temperature rating. This term first appeared in the 1965 NEC cycle. And for many years it simply stated, 'the current-carrying capacity of a conductor expressed in amperes'.

BONDED (BONDING) - Connected to establish electrical continuity and conductivity. It should be noted that the concept of bonding and grounding are not the same. Usually, when proper bonding methods are provided, the bonded equipment is also grounded. But that is not always the case. For example, where metallic objects are installed above 8 feet (2.5m) vertically, or more than 5 feet (1.5m) horizontally from ground or grounded metal objects, the equipment is not required to be grounded (250.110(1)). Equipment installed with these vertical and horizontal clearances may be properly bonded to reduce voltage differences where the equipment may be contacted by people. But, the equipment may not be connected to an equipment grounding conductor.

BONDING CONDUCTOR OR JUMPER - A reliable conductor to ensure the required electrical conductivity between metal parts that are required to be electrically connected.

BONDING JUMPER, EQUIPMENT - The connection between two or more portions of the equipment grounding conductor.

BONDING JUMPER, MAIN - The connection between the grounded (usually the neutral conductor, except on corner-grounded delta systems) and the equipment grounding conductor at the service.

BONDING JUMPER, SYSTEM - The connection between the grounded circuit conductor and the supply-side bonding jumper, or the equipment grounding conductor, or both, at a separately-derived system.

BRANCH-CIRCUIT - The circuit conductors between the final overcurrent device protecting the circuit and the outlet(s).

CIRCUIT BREAKER - A device designed to open and close a circuit by nonautomatic means and to open the circuit automatically on a predetermined overcurrent without damage to itself when properly applied within its rating.

CONTINUOUS LOAD - A load where the maximum current is expected to continue for 3 hours or more. Continuous loads produce additional heating of conductors and equipment. Therefore, it would normally be a requirement to increase conductor sizes and the rating of overcurrent devices to compensate for this additional heat. Also, keep in mind that the ampacity tables, such as NEC Table 310.15(B)(16), identify the continuous-current ratings of conductors, based on the maximum operating temperatures specified in the table. There are four principle determining factors that relate to the conductor's maximum operating temperature. These factors include:

1. Ambient Temperature - which may vary along the conductor length as well as from time to time. The worst case condition must be used to calculate the conductor ampacity.
2. The load current flow through the conductor, which includes fundamental and harmonic currents.
3. The effects of the ambient medium that covers or surrounds the conductors and may impede heat dissipation.
4. The effects of adjacent load-carrying conductors.

For example, the ampacities expressed in NEC Table 310.15(B)(16) are based on an ambient temperature of 30 degrees C., or 86 degrees F., an ambient medium of free air, and no more than 3 current-carrying conductors in a raceway or cable (310.15(B)(3)), with exceptions for certain types of raceways, such as a metal wireway, where the number of current-carrying conductors does not exceed 30 (376.22(B)).

For example, if three 12 AWG copper conductors are installed in a metal conduit in free air with an ambient temperature of 30 degrees C., or 86 degrees F., and the load on each conductor is 20 amperes, the operating temperature of each conductor will reach 60 degrees C., and stabilize at that temperature. So based on these conditions, the conductors will carry 20 amperes of load continuously, with no damage to the insulation with at least a 60°C temperature rating.

However, where branch-circuit or feeder conductors supply continuous loads, these conductors must have an ampacity of 125% of the continuous load, plus any noncontinuous load. An exception would waive this requirement in the event that the overcurrent device is part of an assembly that is listed for operation at 100% of its rating (210.19(A)(1)(a), (215.2(A)(1)(a)).

Also, it should be noted that the concept of continuous loading does not apply to grounded conductors because they normally do not connect to an overcurrent device. An overcurrent device is not to be connected in series with a conductor that is solidly grounded, unless this device opens all of the conductors simultaneously. A common-trip breaker would satisfy this requirement (240.22). If this were done, the continuous loading provision would apply to the grounded conductor, as well as to the ungrounded conductors.

The concept of continuous loading should not be confused with continuous-duty, as associated with typical motor circuits. Continuous-duty is defined in NEC Article 100 as, 'operation at a substantially constant load for an indefinitely long time'. And the Note to Table 430.22(E) states that, 'any motor shall be considered as continuous-duty unless the nature of the apparatus it drives is such that the motor will not operate continuously with load under any condition of use'. The other duty cycles and their conditions of use are expressed in Table 430.22(E). These duty cycles include 'intermittent', 'periodic', 'short-time', and 'varying'. Examples include freight and passenger elevators (intermittent), conveying systems (varying), coal-handling machines (periodic), and motor operated valves (short-time). 430.22 requires that the conductors supplying a single motor for continuous-duty be sized at 125% of the motor full-load current rating. However, this is not taken from the motor nameplate, but from the appropriate tables of Article 430 (430.247, 430.248, 430.249, and 430.250).

Separate motor overload protection is based on the motor nameplate current rating (430.6(A)(2)).

COORDINATION (SELECTIVE) - Localization of an overcurrent condition to restrict outages to the circuit or equipment affected, accomplished by the selection and installation of overcurrent protective devices and their ratings or settings for the full range of available overcurrents, from overload to the maximum available fault current, and the full range of overcurrent protective device opening times associated with those overcurrents. It certainly makes sense to be able to isolate an overload or fault condition and restrict the outage to the affected equipment. And, sometimes this coordination is mandatory, as would be the case for Emergency Systems (700.32). For example, for line and load-side fuses, where the downstream fuse element senses the overcurrent and its element begins to melt, and then arc, and when the arc is extinguished and the fuse is open, the upstream fuse element has not started to melt. Or, where line-side and load-side circuit breakers are to be coordinated, the load-side circuit breaker senses the overcurrent and releases its tripping mechanism and the circuit breaker contacts open and the arc is extinguished before the line-side circuit breaker releases its tripping mechanism. Detailed information on selective coordination is included in Chapter 4.

CURRENT-LIMITING OVERCURRENT PROTECTIVE DEVICE - This is extremely important protection for conductors and equipment. And, it may have an effect on reducing the available fault-current downstream of the protective device. However, care must be exercised when considering this added benefit. For example, when applying the 'Series Ratings' of 240.86, it may not be possible to reduce fault current by the use of a current-limiting overcurrent protective device, and to use this reduction to affect the ratings of downstream equipment. Especially where a significant motor load is supplied (240.86(C)). And it may not be possible to provide 'Selective Coordination' for Series Combination Rated Systems as well. In this respect, this type of protection would not be acceptable where selective coordination is mandatory, such as elevator circuits, where a single feeder supplies more than one elevator car (620.62), and emergency systems (700.32). Current-Limiting Overcurrent Protective Devices have a very fast response time. When operating within their current-limiting range, these devices will interrupt fault-current in ½ cycle, or less, (.008 seconds)(60Hz).

Short-circuit currents reach a peak during the first half-cycle, typically within 60-90 degrees of the voltage-wave. This, in effect, is a worst case condition, as magnetic forces are at their peak and are equivalent to the square of the peak current level. And the thermal energy varies as the square of the RMS current level.

So, if the current-limiting overcurrent protective device opens before the fault current reaches its optimum level, very possibly within ¼ cycle (.004 seconds), the resulting magnetic forces and heat energy will be reduced to manageable levels.

And the use of these devices may have a marked effect on the system equipment grounding conductors. That is, in their ability to provide an 'effective ground-fault current path' (250.4(A)(5)).

DEMAND FACTOR - The ratio of the maximum demand of a system, or part of a system, to the total connected load of a system or part of the system under consideration. In other words, the use of a lower load factor for sizing circuit components where all of the equipment will not be in operation at the same time.

EFFECTIVE GROUND-FAULT CURRENT PATH - An intentionally constructed, low-impedance path, designed and intended to carry current during ground-fault conditions from the point of a ground-fault on a wiring system to the electrical supply source, and that facilitates the operation of the overcurrent protective device or ground-fault detection system. This path may consist of a copper or other corrosion resistant conductor, a metal raceway, the outer covering of a metallic cable assembly, or a metal cable tray. Where properly designed to carry the ground-fault current that may be imposed on it, as well as limiting the voltage-rise on metallic equipment frames, this path will allow the maximum level of ground-fault current to return to the source (transformer, generator, etc.) and cause the rapid clearing of the overcurrent device on a solidly grounded system, or initiating the operation of the ground-fault detection system on a system that is ungrounded or impedance grounded (250.4(A)(5)), (250.4(B)(4)).

ELECTRONICALLY- ACTUATED FUSE - An overcurrent protective device that generally consists of a control module that provides current-sensing, electronically derived time-current characteristics, energy to initiate tripping, and an interrupting module that interrupts current when an overcurrent occurs. Such fuses may or may not operate in a current-limiting fashion, depending on the type of control selected. These devices have similar operating characteristics of a circuit breaker (ANSI/IEEE C37.40).

FEEDER - All circuit conductors between the service equipment, or the source of a separately-derived system, or other power supply source (batteries, solar photovoltaic system, etc.) and the final branch circuit overcurrent device. Feeder conductors do not directly supply utilization equipment.

GROUND - The earth. As a means of clarification, a conducting body may serve in place of the earth as a ground reference point. The metal frame of a car, truck, bus, etc., the metal frame of an airplane, and the metal frame of a portable or vehicle-mounted generator (250.34(A)(B)), are examples of establishing ground references without a physical connection to the earth.

GROUND-FAULT - An unintentional, electrically conductive connection between an ungrounded conductor of an electrical circuit and the normally non-current carrying conductors, metal enclosures, metallic raceways, metallic equipment, or the earth.

GROUNDED (GROUNDING) - Connected to ground or to a conducting body that extends the ground connection.

GROUNDED, SOLIDLY - Connected to ground without inserting any resistor or impedance device. Of course, most systems that are grounded are provided with a solid conductor connection to ground (earth). This provides lightning protection for systems and equipment, as well as a means of stabilizing the voltage-to-ground during normal and abnormal conditions (250.4(A)(1)).

GROUNDED CONDUCTOR - A system or circuit conductor that is intentionally grounded. This includes the grounded conductor of single-phase, 3-wire systems, 3-phase, 4-wire, wye-connected systems, and 3-phase, 4-wire, delta systems, and corner-grounded delta systems.

GROUND-FAULT CIRCUIT INTERRUPTER - A device intended for the protection of personnel that functions to de-energize a circuit or portion thereof within an established period of time when a current to ground exceeds the values established for a Class A device. UL 943 is the testing standard for these devices. The first reference of a GFCI device was in Section 680.4(G), which appeared in 1968, and applied to underwater swimming pool lighting fixtures. These devices had a tripping range of 20mA and were listed as Class B. However, these devices have not been referenced in UL 943 since 2006. A Class A GFCI has a tripping range of 4-6mA. So, no less than 4mA, or 6mA, or higher.

It should be noted that there are 'special purpose' GFCI's that have tripping ranges of 15-20mA. And the system voltages are as follows: Class C-circuit voltage-up to 300 volts-Class D-circuit voltage over 300 volts-Class E-the same as Class D, however the circuit has an equipment grounding system or a system of double insulation.

In addition, 553.4 and 555.3 address the requirement of 'Ground-Fault Protection' for the main overcurrent devices that supply 'Floating Buildings' and 'Marinas'. This protection must be arranged to trip at no more than 100mA (553.4) or 30mA (555.3). This protection should not be confused with the GFCI requirements specified in the NEC.

GROUND-FAULT CURRENT PATH - An electrically conductive path from the point of a ground-fault on a wiring system through normally noncurrent-carrying conductors, equipment, or the earth to the electrical supply source. As a point of reference, during normal or ground-fault conditions, some current returning to the system source will flow through the earth to reach the source. For example, where a grounding electrode is installed at the location of the service equipment and also at the supply transformer, current returning through the grounded (usually neutral) conductor will divide unevenly at the service equipment. This may be normal load-current or ground-fault current. Most of this returning current will flow through the lower impedance grounded conductor to the supply transformer. The remaining current, typically no more than 10% of the total, will return through the earth between the two grounding electrodes. In this regard, the path through the earth parallels the path through the grounded conductor.

GROUND-FAULT PROTECTION OF EQUIPMENT - A system intended to provide protection of equipment from damaging line-to-ground fault currents by operating to cause a disconnecting means to open all ungrounded conductors of the faulted-circuit. This protection is provided at current levels less than those required to protect conductors and equipment from damage through the operation of a supply circuit overcurrent device. As a point of reference, the most common type of fault occurring in distribution systems is a ground-fault. And, by far, the most common type of ground-fault is an arcing-type fault. These types of faults are very difficult to detect, as they strike and restrike intermittently. And, the level of ground-fault current is often too low to cause the overcurrent device to promptly clear (if at all). If allowed to continue, severe damage to equipment may be the result. Sometimes the arcing-fault will develop into a short-circuit, thereby causing the operation of the overcurrent device, but not before significant equipment damage.

GROUNDING CONDUCTOR, EQUIPMENT - The conductive path(s) that provides a ground-fault current path and connects normally noncurrent carrying metal parts of equipment together, and to the system grounded conductor or to the grounding electrode conductor, or both (250.118).

GROUNDING ELECTRODE - A conducting object through which a direct connection to earth is established. There are many types of grounding electrodes, some of which are more effective than others in establishing a connection to the earth and a "0" volts reference point. Ground rods, concrete-encased electrodes, and ground rings are examples of grounding electrodes 250.52(A)(3)(4)(5).

GROUNDING ELECTRODE CONDUCTOR - A conductor used to connect the system grounded conductor (again, usually the neutral) or the equipment grounding conductor to a grounding electrode or to a point on the grounding electrode system. This conductor is not expected to carry significant fault current, although higher frequency lightning currents will flow through this conductor to the earth connection. And, this conductor will not carry significant current for a long period of time. Therefore, its size is limited to approximately 15%–40% of the cross-sectional area of the ungrounded supply conductors. And, its length must be limited to no more than necessary to make the connection to the grounding electrode (250.64)(250.66)(250.166).

INTERRUPTING RATING - The highest current at rated voltage that a device is intended to interrupt under standard test conditions. In order to comply with this form of protection, the available fault-current would be calculated to the line terminals of the fuse or circuit breaker and this would establish the interrupting rating of the device. This would permit the overcurrent device to safely interrupt the fault current. It should be noted that this term also applies to conditions other than short-circuits or ground-faults. For example, the interrupting rating of motor circuit switches or motor controllers, which may be required to interrupt the locked-rotor currents of motors (110.9) (430.83)(A)(1) 430.109(A)(1)).

NEUTRAL CONDUCTOR - The conductor connected to the neutral point of a system that is intended to carry current under normal conditions. It should be noted that the neutral current may include fundamental and harmonic currents. The harmonic currents are caused by nonlinear loads that are supplied by three-phase, four-wire, wye connected systems. In certain applications, the harmonic currents, typically the third harmonic (180 cycle on 60Hz, systems), may cause the neutral currents to exceed the current in the ungrounded conductors.

NEUTRAL POINT - The common point on a wye-connection in a polyphase system, or midpoint of a single-phase, 3-wire system, or midpoint of the single-phase portion of a 3-phase delta system, or, the midpoint of a 3-wire, direct-current system. The neutral point establishes a '0' volts reference point with respect to the earth. However, because the grounding electrode conductor connects the neutral point to the grounding electrode (earth), and there is current flowing through the grounding electrode conductor during normal and fault conditions, the voltage-drop associated with this current flow will cause the neutral point to be at somewhat above '0' volts, with respect to the earth.

NONLINEAR, LOAD - A load where the waveshape of the steady-state current does not follow the waveshape of the applied voltage. Electronic equipment with switching-type power supplies, adjustable-speed drives, and electric-discharge lighting systems are examples of nonlinear loads. Where these types of loads are supplied by 3-phase, 4-wire, wye connected systems, triplen harmonic currents (3rd, 9th, 15th, etc.) will flow in the neutral conductor.

These types of electrical loads produce a condition where the circuit impedance changes with a change in voltage

Excessive heating of the neutral conductor may be the result. Increasing the neutral conductor size, the use of harmonic filters, or the use of zig-zag grounding autotransformers may be necessary where harmonic currents are excessive. However, the use of autotransformers for this purpose is limited in accordance with 215.11 and 450.5.

OVERCURRENT - Any current in excess of the rated current of equipment or the ampacity of a conductor. It may be the result of an overload, short-circuit, or ground-fault. It should be noted that an overcurrent may actually be a normal condition in certain instances. For example, motor starting current, or the energizing current of a transformer or capacitor.

OVERCURRENT PROTECTIVE DEVICE, BRANCH-CIRCUIT - A device that is capable of providing protection for service, feeder, and branch-circuits and equipment over the full-range of overcurrents between its rated current and its interrupting rating. Fuses have a minimum interrupting rating of 10,000 amperes and circuit breakers have a minimum interrupting rating of 5000 amperes (240.60(C)(3)(240.83(C).

OVERCURRENT PROTECTIVE DEVICE, SUPPLEMENTARY - This device provides a limited means of overcurrent protection for specific applications, such as certain types of utilization equipment, including appliances and luminaires. As the name implies, these overcurrent devices supplement, and are not used in lieu of the branch-circuit overcurrent devices, as they do not have a listed interrupting rating. In most cases, the supplementary overcurrent device is short-circuit tested on the load side of a branch circuit overcurrent device (UL 1077).

RACEWAY - An enclosed channel of metallic or nonmetallic materials designed expressly for holding wires, cables, or busbars, with additional functions as permitted in this Code. In addition to providing physical protection for the conductors, a metallic raceway is typically used for equipment grounding functions (250.118). It should be noted that a cable tray (Article 392), whether open or covered, is not an example of a raceway.

SEPARATELY-DERIVED SYSTEM - An electrical source, other than a service, having no direct connection(s) to circuit conductors of any other electrical source, other than those established by grounding and bonding connections .

SHORT-CIRCUIT CURRENT RATING - The prospective symmetrical fault current at a nominal voltage to which an apparatus or system is able to be connected without sustaining damage exceeding defined acceptance criteria. Modular data centers and surge-protective devices are examples of equipment that require overcurrent protection in accordance with their short-circuit current rating. This means that specific calculations must be made to determine the available fault-current in a system, and the appropriate overcurrent protection must be provided to protect the equipment in accordance with the short-circuit current rating of the equipment (646.7), (285.7).

SUPERVISED INDUSTRIAL INSTALLATIONS - This term has been used for many years as a means of qualifying certain conditions that would not be otherwise acceptable. For example, applications of 'feeder taps' and 'transformer secondary conductors of separately-derived systems' (240.92(B), (C)). In these cases, short-circuit and ground-fault protection for conductors may be modified because of qualified maintenance and supervision. The normal rules covering tap conductor short-circuit current ratings may be determined in accordance with Table 240.94(B), as opposed to the normal method of 240.21(B). However, this provision does not negate the fact that short-circuit and ground-fault protection for these conductors is required.

TAP CONDUCTORS - In this respect, conductors are permitted to be provided with overcurrent protection in excess of their normal ampacity where certain conditions are met. But, there have been many abuses to the various 'tap rules'. So, let's analyze these rules and recognize some of the hazards. This is a relatively long summary, so carefully read and understand this material.

1. **Section 210.19(A)(3)(A)(4)** - For household ranges and cooking ppliances, conductors with an ampacity of at least 20 amperes may be tapped to a 50 ampere branch circuit. In a typical dwelling occupancy, the available short-circuit current is limited, and very often does not exceed 6000 amperes. So, using this tap rule may not be a problem.

However, the fault-current calculations must still be done to assure that proper protection has not been compromised.

2. **Section 210.19(A)(4), Exception No.** 1 permits conductors, with an ampere rating of 15 or 20 amperes to be protected at 40 or 50 amperes for certain types of equipment, such as lampholders or luminaires where the tap conductors are limited to 18 inches (450mm) in length. Or, tap conductors from a luminaire which extend to an outlet box which is at least 1 foot (300 mm) from the luminaire (to protect the conductors from excessive heat). These conductors may be 14 AWG (2.08mm^2) and up to 6 feet (1.8m) in length. And they must be provided with physical protection as well. Or, tap conductors, not smaller than 14 AWG (2.08mm^2) may supply individual outlets (such as lighting outlets, but not receptacle outlets), where the length is limited to 18 inches (450mm). Tap conductors for infrared lamp industrial heating appliances may be 14 AWG (2.08mm^2). Or, the nonheating leads of deicing and snow-melting cables and mats may be 14 AWG (2.08mm^2). In each of these rules, the tap conductors must not be smaller than 14 AWG (2.08mm^2), and must have an ampacity that is sufficient for the connected load.
3. **Section 240.5(B)(2)** - Fixture wire may be tapped to branch-circuit conductors protected at 20 amperes if the conductors are 18 AWG (0.823mm^2), and no longer than 50 feet (15m), or 16 AWG ((1.31mm^2), and no longer than 100 feet is (30m).
4. **Section 240.21** - This section requires a detailed analysis, as it involves feeder taps and transformer applications.
   240.21(B)(1)-Feeder Taps - Where these tap conductors are not more than 10 feet (3m) in length and meet the following conditions:
   a. The ampacity of the taps conductors is not less than the load to be served.
   b. The tap conductor ampacity is not less than the rating of the equipment that contains an overcurrent device supplied by the tap conductors. This is a modification of the 2014 NEC, in that prior to this Code cycle, the equipment supplied was not specifically required to have an overcurrent device, only that the equipment had an ampere rating equal to, or greater than, the tap conductors.
   c. The tap conductors have an ampacity not less than the rating of the overcurrent device at the termination of the tap conductors.
   d. Once the termination is made, the tap conductors do not extend beyond the equipment they supply.
   e. The tap conductors are provided with physical protection in the form of a raceway.
   f. For field installations, where the tap conductors leave the enclosure or vault where the tap is made, the ampacity of the tap conductors

is at least one-tenth (1/10) of the rating of the overcurrent device protecting the feeder conductors. Therefore, if a 10 AWG (5.26mm^2) THHN-copper conductor (40 amperes-Table 310.15(B)(16)), is tapped to a larger feeder conductor, the upstream overcurrent protection may be as large as 400 amperes.

Well, let's look at this example. If the tap is made relatively close to the location of the 400 ampere overcurrent device, and the available fault-current is calculated to the point on the feeder where the tap is made, and the fault-clearing time of the 400 ampere overcurrent device is one-cycle (.016 seconds), the fault-current at the tap could not exceed 4349 amperes, which is the one-cycle insulation withstand rating of the 10 AWG copper conductor, based on the 75 degree C. insulation rating of this conductor. In this example, we are using a 10 AWG copper conductor with THHN (90 degree C.) insulation. But, this conductor will be terminated into terminals that have a lower temperature limit, which is 60 degrees C., or 75 degrees C., based on either the wire size or the rating of the circuit (110.14(C)(1)(a)). I am stressing the issue of simply using this tap rule without considering other factors which may affect its use. If the calculated available fault-current at the point where the tap is made exceeds 4349 amperes, the tap conductor size would have to be increased accordingly. Or, the available fault-current would have to be reduced, possibly through the use of a current-limiting overcurrent protective device. It should be noted that the tap rules expressed in Article 240 (Chapter 2) cannot be used to amend the other rules in Chapters 1 through 4. In our example of the 10 foot tap rule, sizing the tap conductors at 10% of the ampere rating of the upstream overcurrent device, and possibly compromising the short-circuit withstand rating of the tap conductor insulation is a violation of Section 110.10. Section 90.3 states that Chapters 1 through 4 are general in nature. And amendments to these general provisions may appear in Chapters 5, 6, and 7, which may supplement or modify the requirements in Chapters 1-7.

5. **Taps not over 25 feet (7.5m) long** - This tap rule is somewhat more reasonable in application due to the larger size of conductor. But this does not mean that we can overlook other rules in the process. So, here it is.
   a. The ampacity of the tap conductor must be at least 1/3 of the rating of the overcurrent device protecting the feeder conductors.
   b. The tap conductors terminate in a single circuit breaker or single set of fuses that will limit the load to the ampacity of the tap conductors.

c. The tap conductors are protected from physical damage by being enclosed in an approved raceway or other approved means.

This tap rule has many common applications and it can be seen that with a protective device with a rating of three times the ampacity of the tap conductors, it is less likely that short-circuit or ground-fault protection for the smaller conductor will be undermined. For example, consider a 3-phase feeder, consisting of a 350 ampere circuit breaker which is protecting 500 kcmil (253mm^2) THHN copper conductors. The load is continuous, and the normal ampacity of these conductors is 430 amperes, assuming that there are no additional correction or adjustment factors applied, such as ambient temperatures exceeding 30 degrees C. or 86 degrees F., and/or proximity effects (the effects of adjacent current-carrying conductors). Due to the continuous load, the maximum load permitted on the 350 ampere circuit breaker is limited to 280 amperes (80% of 350 amperes), (215.3). And the feeder conductors must have an ampacity of 125% of the continuous load, or 350 amperes (1.25 × 280 amperes) (215.2(A)(1)).

Finally, the 350 ampere circuit breaker has terminals which have a temperature limit of 75 degree C., (110.14(C)(1)(b)). And the ampacity of the 500 kcmil ((253mm^2) copper conductors is 380 amperes at 75° C., (Table 310.15(B)(16)). And now, for the tap conductors. Their ampacity must be at least one-third of the ampere rating of the 350 ampere circuit breaker, or 117 amperes. Once again, if we use THHN insulated copper conductors, we may use a 1 AWG (42.41mm^2) conductor, which has a normal ampacity of 145 amperes at 90 degrees C. At 75 degrees C., this conductor has an ampacity of 130 amperes. Of course, 130 amperes is not a standard size overcurrent device (240.6(A)). And the tap rule specifically states that these tap conductors must terminate in a circuit breaker or single set of fuses that will limit the load to the ampacity of the tap conductors. Permission to use the next standard size of overcurrent device (240.4(B)) is not given. These tap conductors would terminate into a 125 ampere overcurrent device. And the 75 degree C. ampacity of the 1 AWG (42.41mm^2) copper conductors is 130 amperes, so the provisions of 110.14(C)(1)(b) are satisfied.

This is a relatively easy process if you make the connection to each of these requirements. But, even so, the fault-current calculations must be done to assure that the circuit components are protected in accordance with their short-circuit current ratings (110.10).

And the available short-circuit current at the line terminals of the fuse or circuit breaker at the termination of the tap conductors will establish its interrupting rating (110.9).

6. **Section 240.21(B)(3)** - This tap rule involves a transformer, where the total length of one primary, plus one secondary conductor, excluding any primary conductor that is protected at its ampacity, is not longer than 25 feet (7.5m).
   a. The conductors supplying the primary of the transformer have an ampacity of at least one-third of the ampere rating of the overcurrent device protecting the feeder conductors.
   b. The secondary conductors have an ampacity that is not less than the value of the primary-to-secondary transformer voltage ratio multiplied by one-third of the rating of the overcurrent device protecting the feeder conductors.
   c. The total length of one primary plus one secondary conductor does not exceed 25 feet (7.5m).
   d. The primary and secondary conductors are provided with physical protection, such as a raceway or other approved means.
   e. The secondary conductors terminate in a circuit breaker or set of fuses that limit the load current to no more than the conductor ampacity. Once again, if the secondary conductor ampacity does not correspond to the standard ampere ratings of overcurrent devices in 240.6(A), the use of the next standard size is not permitted.

Example

Feeder Protection – 300 Amperes
Transformer..........75 kVA-3-Phase

$$\frac{75{,}000\,VA}{480V \times 1.732} = 90.21\,Amperes$$

Primary Overcurrent device = 100 amperes (1/3 of 300 amperes)
Full-Load Primary Current..............90 amperes
Full-Load Secondary Current..........208 amperes
Voltage...........Primary.......... 480 Volts
Voltage...........Secondary ..... 208/120 Volts
Primary-to-Secondary Voltage Ratio - 480/208 = 2.31
100 amperes × 2.31 = 231 amperes

This establishes the ampere rating of the secondary conductors.
Primary Conductors ....3 AWG (26.67mm^2) THHN
Copper-115 Amperes (90°C)
100 amperes (60°C.)

Secondary Conductors ....250 kcmil THHN-Copper-
290 amperes-90°C.
255 amperes-75°C

The secondary conductors terminate into a 225 ampere circuit breaker or fuse. The terminal rating of the circuit breaker or fusible switch is 75°C. The secondary conductor ampacity at 75°C. is 255 amperes. Once again, the use of the next standard size circuit breaker or fuse (250 amperes) is prohibited.

The primary conductors are tapped to a feeder, which is protected at 300 amperes.

It should be noted that other NEC sections may be applicable when using this tap rule. For example, where the secondary conductors terminate into a panelboard, the ampere rating of the panelboard must not be less than the feeder conductor ampere rating (408.3(D)). And the panelboard main overcurrent device in this example is 225 amperes (408.36).

Because a transformer is used in this tap rule, the overcurrent protection rules for transformers (450.3(B)) must be satisfied as well.

At 75kVA, and a 480 volt primary, the transformer primary overcurrent protection rules for transformers (450.3(B)) must be satisfied.

At 75kVA, and a 480 volt primary, the transformer primary overcurrent device may be rated at 250% of the full-load primary current rating (450.3(B)) (90 amperes × 2.5 = 225 amperes)

By simply using the transformer tap rule, we deduced that the primary conductors may be tapped to a feeder which is protected at 300 amperes.

But 450.3(B) only permits the 75kVA, 480 volt transformer to be protected at 225 amperes (250% of the full-load primary current of 90 amperes).

So, the primary feeder conductors may not be protected at 300 amperes, as it appears in the tap rule of 240.21(B)(3), because 450.(B) restricts the primary overcurrent device to no more than 225 amperes (250% of 90 amperes).

By reducing the feeder overcurrent device to 225 amperes, this would also affect the size of the secondary conductors. Now, the calculation would be: 225 amperes × .7666 (primary-to-secondary voltage ratio (2.3 × $^1/_3$) = 172 amperes. Using a THHN insulated copper conductor, a 2/0 (67.44mm^2) conductor may be used (195 amperes at 90°C.). And this conductor has a 75°C. ampacity of 175 amperes. And 175 amperes is a standard size overcurrent device (240.6(A)).

Once again, reverting to Section 90.3, a reference in Chapter 2 (240.21(B)(3)) cannot amend a reference in Chapter 4 (450.3(B)).

The secondary overcurrent device in this example is 225 amperes. And 450.3(B) permits a secondary overcurrent device to be rated or set at 125%

(1.25 times) of the full-load secondary current of the transformer. Which, in this case, would be 260 amperes (208 amperes × 1.25).

However, Section 240.21(B)(3)(5) specifically states that the secondary conductors terminate in a single circuit breaker or set of fuses that will limit the load current to no more than the conductor ampacity that is permitted by 310.15. In effect, the ampacity, as referenced in Table 310.15(B)(16), with any additional factors, such as possible ambient temperature correction. In our example, the secondary conductors are 250 kcmil THHN copper (290 amperes@ 90 degrees C.). But, as we alluded to earlier, the ampacity of this conductor is 255 amperes in order to comply with Section 110.14(C)(1)(b), as it relates to the 75 degree C. temperature limit of the terminals of the secondary circuit breaker or fusible switch.

Finally, the end result is that the primary overcurrent device appears to be acceptable at a rating of 300 amperes, according to Section 240.21(B)(3). But because of Section 450.3(B), the maximum rating of the overcurrent protection for the primary of this transformer is 250% of the transformer primary full-load current rating. That is, 90 amperes × 2.5 = 225 amperes. So, we determined that the maximum size of the feeder overcurrent device is 225 amperes. And based on this rating, the secondary conductors may have an ampacity of 175 amperes, or 2/0 copper 67.44mm^2) at 75°C.

Once again, this seems like a lesson in futility, but the requirements must work in harmony, so that, by complying with one Code section, you do not violate the provisions of another section(s).

7. **Taps over 25 feet (7.5m) long -** This is a limited use tap rule, in that, it is permitted to be used only in 'high bay manufacturing buildings', where the height at the walls is over 35 feet (11m). These provisions include the following:
   a. Maintenance and supervision must ensure that only qualified persons service the systems.
   b. The tap conductors may not extend more than 25 feet (7.5m) in a horizontal plane (total). So, possibly, 15 feet (4.57m) and 10 feet (3.05m), or 25 feet (7.5m) in one continuous length. And the total horizontal and vertical length is not to exceed 100 feet (30m).
   c. The tap conductor ampacity is not less than 1/3 of the rating of the overcurrent device protecting the feeder conductors.
   d. The tap conductors terminate at a single circuit breaker or single set of fuses that will limit the load to the ampacity of the tap conductors.
   e. The tap conductors are provided with physical protection in the form of a raceway, or other approved means.

f. The tap conductors are not spliced.
g. The minimum size of the tap conductors is 6 AWG 13.3mm^2) copper or 4 AWG (21.15 mm^2) aluminum.
h. The tap conductors do not extend through walls, ceilings, or floors.
i. The tap is made at least 30 feet (9m) above the floor.

Once again, these tap conductors may have an ampacity that is significantly less than the ampere rating of the upstream feeder overcurrent device. Which is acceptable, if this overcurrent device provides the necessary short-circuit and ground-fault protection for these tap conductors.

8. **Outside Taps of Unlimited Length** - These conductors are located outside of the building or structure, except at the point of termination. The following conditions apply:
   a. The tap conductors are protected from physical damage. These conductors would very often be run underground. So, the installation in a metallic or nonmetallic conduit would be appropriate. And if installed overhead, with the proper elevation in accordance with 225.18, the physical protection is inherent to the conductor elevation.
   b. The tap conductors terminate in a single circuit breaker or set of fuses that will limit the load to the ampacity of the tap conductors.
   c. The overcurrent device for the tap conductors is an integral part of the disconnecting means, or located immediately adjacent to the disconnecting means.
   d. The disconnecting means is readily accessible and its location is:

Outside of the building or structure, or, inside, nearest the point of entrance of the tap conductors.

If installed in accordance with 225.32–230.6 (conductors considered outside the building), nearest the point of entrance of the tap conductors.

In this example, these conductors are tapped to another set of conductors outside of the building or structure and extended any length and terminated into an overcurrent device which is located either inside or outside of the building or structure. And this overcurrent device has a rating that does not exceed the ampacity of the tap conductors. It should be noted that this overcurrent device only provides overload protection (230.90) and not overcurrent protection, that is, short-circuit and ground-fault protection.

Once again, this tap rule does not mitigate the need to provide short-circuit and ground-fault protection for these tap conductors.

9. Transformer Secondary Conductors - Transformer secondary conductors feeding a single load may be extended from the transformer secondary terminals without overcurrent protection as specified in 240.21(C)(1) through (C)(6). However, the use of the next larger overcurrent device where the secondary conductor ampacity does not correspond to the standard ampere rating of an overcurrent device (240.4(B)), is not permissible.
10. Protection by Primary Overcurrent Device - Generally, when dealing with a single-phase transformer with a two-wire (single-voltage) secondary, or a three-phase, three-wire, delta-to-delta connected transformer, it is permissible to use the primary overcurrent device as a means of protecting the secondary conductors. After sizing the primary overcurrent device in accordance with 450.3, the secondary conductor ampacity is based on the secondary-to-primary transformer voltage ratio.

For example, a 25kVA, single-phase, 480 volt primary to 240 volt (single-voltage) secondary has a full-load primary current of 52 amperes. Table 450.3(B) permits the primary overcurrent device to be rated or set at 65 amperes. Exception No. 1 permits this overcurrent device to be the next standard size, because 65 amperes is not a standard size (240.6(A)). The next standard size is 70 amperes.

The secondary-to-primary transformer voltage ration is 240/480 = ½ , or .5.

Primary overcurrent device - 70 amperes /.5 = 140 amperes.

Therefore, if the secondary conductor ampacity is at least 140 amperes, these conductors are considered to be protected by the primary (70 ampere) overcurrent device.

This also applies to three-phase, three-wire (single-voltage) secondary systems.

For single-phase, three-wire, secondary systems with dual voltages (nominal 240/120 volts), the secondary conductors require overcurrent protection in accordance with 240.21(C).

And for three-phase, four-wire, delta-to-delta systems, or delta-to-wye systems, the secondary conductors must be protected, irrespective of the primary overcurrent device in accordance with 240.21(C).

In a delta-to-wye system, there is a phase-shift of 30 degrees. The primary (delta) voltage will lead the secondary (wye) voltage by 30 degrees. Consider the phase-angle on the delta primary to be: Voltage Phases A-B at '0' degrees, Voltage Phases B-C at -120 degrees, and Voltage Phases C-A at -240 degrees. The wye secondary would be at Voltage A-N at '0' degrees, Voltage B-N at -120 degrees, and Voltage C-N at 240 degrees. So, the secondary voltage would be - Voltage A-N minus Voltage B-N = Voltage A-B at 30 degrees.

Based on this phase-shift, the current in the secondary is not in direct proportion to the current in the primary windings. This applies to normal operating conditions, as well as short-circuit and ground-fault conditions.

Couple this condition with the fact that the primary overcurrent device may be rated at 250% of the transformer primary current rating (Table 450.3(B)), and we can see the need for the secondary conductor overcurrent device.

11. Transformer secondary conductors not over 10 feet (3m) long. These secondary conductors connect directly to the transformer secondary terminals and extend a limited length to the connected equipment as follows:
    (Note: This explanation and associated examples is quite long and detailed. But, this information may be used for other installations, and not just for this tap rule.)
    a. The ampacity of the secondary conductors must not be less than the calculated load.
    b. The ampacity of the secondary conductors must not be less than the rating of the equipment which contains an overcurrent device (such as a main circuit breaker in a panelboard). Or, not less than the rating of the overcurrent device at the termination of the secondary conductors. In any case, there will be an overcurrent device at the termination of these secondary conductors.
    c. The secondary conductors do not extend beyond the equipment that they supply.
    d. The secondary conductors are provided with physical protection in the form of a raceway or other approved means.
    e. For field installations, where the secondary conductors leave the enclosure or vault where the connection is made, the rating of the overcurrent device protecting the primary of the transformer, when multiplied by the primary-to-secondary transformer voltage ratio, is not more than 10 times the ampacity of the secondary conductors.

Example

37.5kVa-3-Phase Transformer - 480/208/120 Volts
Full-load Primary Current - 45 Amperes
Primary Overcurrent Device Rating - 56 Amperes (45 amperes × 1.25) (450.3(B), Note 1). The next standard size is 60 amperes.
1/10 (10%) of the rating of the primary overcurrent device - 6 amperes

Primary-to-Secondary Transformer Voltage Ratio - 480/208 = 2.31
Minimum Ampacity of Secondary Conductors - 6 amperes × 2.31 = 13.86, or 14 amperes = 14 AWG-Copper (240.4(D)).

In this example, the transformer primary conductors would have an ampacity of 56 amperes, or possibly, 6 AWG-XHHW-copper.

The secondary conductors are 14 AWG-copper.

And yet, the rating of the primary overcurrent device is 60 amperes. We have already discussed the 30 degree phase-shift from primary-to-secondary in this delta-to-wye connection. Would the primary overcurrent device (60 amperes) protect the secondary-14 AWG copper conductors?

This is certainly questionable, and must be a consideration in applying this tap rule.

Example

Transformer – 37.5kVa-480 Volts-208/120 Volts
Percent Impedance - 1.5% (Z)
Full-load Secondary Current - 104 amperes (37,500VA/208 × 1.732)
100/1.5 (Z) = 66.67
104 amperes × 66.67 = 6934 amperes

This is the short-circuit current available at the transformer secondary terminals. There is no motor contribution in this example. If there were, the total motor load would be multiplied by 4 and added to the calculated short-circuit current. The insulation withstand rating of the 14 AWG- copper conductors (75 degrees C.) for 1 cycle (.016 seconds) = 1715 amperes, which is the fault clearing time of the primary overcurrent device. With 6934 amperes available at the point of connection of the 14 AWG (2.08mm^2) copper conductors, and an insulation withstand rating of 1715 amperes for 1 cycle, and the only overcurrent protection for these conductors is the 60 ampere primary overcurrent device, it is extremely doubtful that the 14 AWG copper conductors are afforded the protection required by Section 110.10.

As stated previously, due to the fact that this transformer is connected delta-to-wye, there is a 30 degree phase-shift from primary-to-secondary. And, a short-circuit or ground-fault in the secondary conductors, from the terminals of the transformer to the location of the secondary overcurrent device, may not cause the primary overcurrent device to promptly clear (if at all).

Some people will say that such a fault is very unlikely, due to the limited length of the secondary conductors (10 feet - 3m). And the fact that physical protection, in the form of a raceway is required for the secondary conductors. But this is still a possibility. And this fault may not be in the secondary conductors, but within the transformer secondary windings, where a secondary overcurrent device would be of no help.

We calculated the secondary fault current at 6934 amperes, based on the transformer rating of 37.5kVA and a percent impedance (Z) of 1.5%. It should be noted that UL 1561 permits listed transformers of 25 kVA and larger to have a plus or minus 10% impedance from the value specified on the nameplate, which in our example would be either 1.65% or 1.35%. If we use the worst case condition, the available short-circuit current at the transformer secondary terminals would increase from 6934 amperes to 7703 amperes.

1.50 x .9 = 1.35 (Z)

100/1.35 = 74.07   104 (Full-load Secondary Current)

$\times$ 74.07

7703 amperes (Available Short-Circuit Current at Transformer Secondary Terminals)

What are the possible options?

The first is to increase the size of the secondary conductors.

We made an assumption that a fault in the secondary conductors would clear the primary 60 ampere fuses in one cycle (.016 seconds). This rapid clearing is probably not the case, even if the primary overcurrent device was reduced in size. So, a determination must be made before we begin our analysis.

The minimum size of our secondary conductors was determined to be 14 AWG (2.08mm^2) copper. But with an available fault-current of 6934 amperes, or a worst case of 7703 amperes, these secondary conductors, with a 1 cycle conductor insulation withstand rating of 1715 amperes, would not be acceptable.

The obvious solution would be to increase the conductor size.

How about the use of 12 AWG (3.31mm^2)? – 12 AWG = 6530 circular mils

6530 circular mils/42.25 = 155 amperes (Note: One ampere for every 42.25 circular mils of conductor cross-sectional area for 5 seconds.)

155 amperes $\times$ 155 amperes $\times$ 5 seconds = 120,125 ampere-squared seconds

120,125 ampere-squared seconds/.016 seconds (1 cycle) = 7,507,813

The square-root of 7,507,813 = 2740 amperes for 1 cycle

This a little better, but not nearly enough

Let's try 10 AWG (5.26mm^2) copper - 10 AWG = 10,380 circular mils

10,380 circular mils/42.25 = 246 amperes (5 seconds)

246 amperes $\times$ 246 amperes $\times$ 5 seconds = 302,580 amperes-squared seconds

302,580/.016 (1 cycle)=18,911,250

The square-root of 18,911,250 = 4349 amperes for 1 cycle
Again, a little better, but still not enough
Now let's try 8 AWG (8.37mm^2) copper - 8 AWG = 16,510 circular mils
16,510 circular mils/42.25 = 391 amperes (5 seconds)
391 × 391 × 5 seconds = 764,405 ampere-squared seconds
764,405/.016 (1 cycle) = 47,775,313
The square-root of 47,775,313 = 6912 amperes for 1 cycle
If we used the original fault-current of 6934 amperes at the transformer secondary terminals, and not the worst case of 7703 amperes, we would almost be there (6912 vs. 6934).
Of course, we want to try the next wire size. It would be as follows:
6 AWG (13.3mm^2) copper - 6 AWG = 26,240 circular mils
26,240 circular mils/42.25 = 621 amperes (5 seconds)
621 amperes × 621 amperes × 5 seconds = 1,928,205 ampere-squared seconds
1,928,205/.016 (1 cycle) = 120,512,813
The square-root of 120,512,813 = 10,978 amperes for 1 cycle
And finally, we have the proper wire size for this example (6 AWG (13.3mm^2) copper
And, don't forget that the primary conductors were 6 AWG-XHHW copper
This exercise may seem to be extreme. But, based on the conditions that were specified, and these are normal conditions, there is no other way. Section 110.10 is all inclusive, in that short- circuit and ground-fault protection for conductors and equipment cannot be compromised.

Yes, by simply reading and applying the provisions of the tap rule, it appears that the secondary conductors could be as small as 14 AWG (2.08mm^2) copper. But, based on the transformer size and its impedance of 1.5%, we found the fault current to be higher than the short-circuit current withstand rating of the conductor insulation, based on the one-cycle clearing time of the transformer primary overcurrent device.

Of course, the transformer impedance may be higher, let's say 2.5%

100/2.5=40 104 amperes (Full-load Secondary Amperes)
× 40
4160 amperes (Available Short-Circuit at Transformer Secondary Terminals)

100/2.5 (worst case Z @ 0.9 of 2.5) = 44.44 104 amperes (FLSA)
× 44.44
4622 amperes (Worse Case)

Based on these conditions, we may be able to use a 10 AWG (5.26mm^2) copper conductor. The one-cycle insulation withstand rating of this conductor is 4349 amperes, which is midway between the normal and worst case fault-current calculations of 4160 and 4622 amperes at the transformer secondary terminals.

And, if the fault occurred several feet, or possibly at the end of the 10 foot (3m) length of these conductors, the fault-current will be reduced to less than these values.

For example, at the end of the 10 foot (3m) length, based on the use of 10 AWG (5.26mm^2) copper conductors, the available fault-current would be 3074 amperes, or 3416 amperes (worst case).

Note: Calculations made from Eaton Electrical Protection Handbook-SPD

And finally, the operating characteristics of the primary overcurrent device must be known in order to determine the fault-clearing time of this device. We have, hypothetically, assumed a clearing time of one-cycle, or .016 seconds.

This detailed analysis is not a lesson in futility. It is provided as a means to consider each aspect of the NEC provisions, and not just concentrating on one specific section. But, correlating the appropriate and relative Code sections in such a way that, by complying with one section, we do not overlook other relative sections that may cause violations of these rules. And, in so many cases, this is so easy to do.

12. Where there are supervised maintenance operations, and conductors supply switchgear or switchboards in industrial installations, and the length of the transformer secondary conductors do not exceed 25 feet (7.5m), the following conditions apply:
    a. The conditions of maintenance and supervision assure that only qualified persons service the installation.
    b. The ampacity of the secondary conductors is not less than the secondary current rating of the transformer, and the rating of the overcurrent device, or the sum of the ratings of multiple devices, do not exceed the secondary conductor ampacity.
    c. If multiple overcurrent devices are used, they must be grouped in the same location.
    d. The secondary conductors are provided with physical protection in the form of a raceway or other approved method of protection.

    This rule is easy to follow. And the secondary conductors do have a minimum size in relation to the secondary current rating of the transformer. So, depending on the type of system (single or three-phase), and the rating of the transformer primary overcurrent protection, and its type, we may determine whether the secondary conductors have the proper overcurrent protection.

13. Outside Secondary Conductors - This is a relatively common application where the tap conductors are connected to a set of feeder conductors outside of a building or structure, and extended any length to the building or structure where they terminate at a fusible switch or circuit breaker. This device is outside or immediately inside and nearest the point of entry of the conductors in a readily accessible location. The 'Authority Having Jurisdiction' will determine how long the inside conductors may be before they reach the fusible switch or circuit breaker. The ampere rating of the fuse or circuit breaker is to be no larger than the ampacity of the supply conductors.

The conductors that are tapped from these building supply conductors may be larger or smaller than the feeder conductors. However, if they are smaller, care must be exercised to be sure that the conductor short-circuit protection has not been compromised.

These tap conductors are required to have physical protection, very likely in the form of an approved raceway, or isolated by elevation in accordance with 225.18.

14. Secondary Conductors from a Feeder Tapped Transformer - This section reverts to 240.21(B)(3), and the protection permitted by this tap rule.
15. Where the Secondary Conductors Not Over 25 Feet Long and the following apply:
    a. The secondary conductors must have an ampacity that is not less than the primary-to- secondary voltage ratio multiplied by one-third of the rating of the overcurrent device protecting the primary of the transformer.

Example

Single-Phase-50 kVA Transformer-480 Volt/240 Volt
Primary-to-Secondary Voltage Ratio-480 Volt/240 Volt = 2/1, or 2
Transformer Primary Overcurrent Device - 50,000VA/480 Volts = 104 amperes
104 amperes × 1.25=130 amperes (450.3(B)) and Note 1 of this Table) - 150 amperes (next standard size from 240.6(A)
One-third of 150 amperes = 50 amperes
50 amperes x 2 (480v/240v = 2/1, or 2) = 100 amperes
The secondary conductors have an ampacity of 100 amperes, or more.

   b. The secondary conductors terminate in a single circuit breaker or set of fuses that will limit the load current to no more than permitted by 310.15.
   c. In this case, a 100 ampere rated overcurrent device. The conductors must have an ampacity of 100 amperes (1 AWG Copper) (42.40mm^2).

   d. The ampacity is based on the 60°C. ampacity of the conductor (110.14)(C)(1)(a)(1)).
   e. The secondary conductors have suitable physical protection, such as an approved raceway or other approved means.
16. Busway Taps- A busway tap rule that is permitted for industrial establishments (368.17(B), Exception) allows a smaller rated busway to be tapped to a larger rated busway without overcurrent protection at the tap. The smaller busway must be at least one-third of the ampere rating or setting of the overcurrent device ahead of the smaller busway and the smaller busway is not to be in contact with combustible material.
17. Motor Circuit Taps - For motor feeder taps, the normal tap rules of 240.21(B) apply with regard to length (10 feet, 25 feet, etc.) and upstream overcurrent protection as follows:
   a. Where the tap conductors are limited to 10 feet (3m) in length and the up stream overcurrent protection is no more than 1000 percent (10 times) of the tap conductor ampacity. Physical protection (raceway, or other approved means) must be provided for the tap conductors
   b. Or, the tap conductors may be 25 feet (7.5m) in length and have an ampacity of at least one-third of the ampacity of the feeder conductors and be provided with suitable protection from physical damage.
   c. Or, the tap conductors have the same ampacity as the feeder conductors.
   d. An exception permits the use of the 100 foot (30m) tap rule for high-bay manufacturing buildings (240.21(B)(4)).

   For motor branch-circuit taps, Section 430.53 applies as follows:
   a. Several motors, not exceeding one horsepower, may be supplied from a 120 volt branch-circuit protected at no more than 20 amperes.
   Or, a branch-circuit of 1000 volts or less protected at 15 amperes, providing that the full-load current rating of each motor is not more than 6 amperes, the rating of the fuse or circuit breaker does not exceed the marked rating on any of the controllers, and individual thermal overload protection is provided in accordance with 430.32. If the branch-circuit protection is sized to protect the smallest rated motor, two or more motors, or one or more motors and other loads may be on the same branch-circuit, providing the branch-circuit overcurrent device will not open under the most severe conditions encountered. Once again, each motor must have individual overload protection.
   b. For other group installations, one branch-circuit is permitted to supply two or more motors where the motor controllers and overload

devices are installed in a listed factory assembly, and the overcurrent devices are part of the listed assembly or specifically marked on the assembly with regard to their rating.

c. Where the overcurrent devices, controller, and overload devices are installed in the field in separate assemblies that are listed for the purpose and provided with manufacturer's instructions for use as a group with the following instructions:
   1) The overload devices for each motor are listed for group installations with the maximum size fuse or circuit breaker specified for the overload devices.
   2) Each controller is listed for group installation with a maximum rating of fuse or circuit breaker (or both). Note: If a specific type of device is specified, only that type of device may be used in order to comply with the listing instructions (110.3(B)).
   3) Circuit breakers must be listed and of the inverse-time type.
   4) The branch-circuit overcurrent protection has a rating that does not exceed that specified in 430.52 for the highest rated motor, plus an amount equal to the sum of the other motors and other loads connected to the circuit.
   5) The overcurrent protection for the branch-circuit is not larger than that permitted for the overload relay protecting the smallest motor of the group.
   6) For nonmotor loads, the overcurrent protection must comply with the applicable provisions of Article 240.
   7) For single motor taps, the general rules for 10 and 25 foot (3m and 7.5m) taps apply. However, if manual motor controllers are used, they must be marked "Suitable for Tap Conductor Protection in Group Installations."

18. Battery Conductors - Battery conductors may extend to overcurrent devices where these device are as close as practicable to the battery terminals.

This ends our analysis of feeder taps. This is a long summary of a concept that permits connected equipment to be supplied from circuits that may be conveniently located and that may have the required capacity to supply additional equipment, without the need to extend conductors for greater distances. This may be a cost effective approach at times, and as long as all factors are considered, there is an advantage to the use of these tap rules.

2

Conductor Insulation Withstand Ratings, Terminal Withstand Ratings, Conductor Fusing Currents, and Overcurrent Protection Over 1000 Volts

Article 240 is devoted to provide the general requirements relating to overcurrent protection and the devices which provide protection against these conditions in accordance with the definition of the term 'Overcurrent' and related terms as defined in Article 100.

Certainly, a major concern is excessive heat and the damage associated with thermal stress to various circuit components.

I can only imagine the challenges that were faced by the early users of electrical power systems. Primarily, buildings were made of combustible wood construction. As an electrician working in New Jersey, I became aware of knob-and-tube wiring (Article 394). Especially in dwelling occupancies, but in some commercial installations as well. But, surprisingly, in many case, the wiring was in relatively good shape. Even in attics, where the ambient temperature was, at times, excessive. Of course, in the dwelling occupancies, the electrical loads were typically intermittent, and for the most part, without the significant appliance loads of today.

However, in many commercial establishments, where the electrical load, especially lighting, was continuous, the wiring was often in poor condition.

Knob-and-tube wiring is only permitted for extensions in existing systems and by 'special permission' (the written consent of the Authority Having Jurisdiction (394.10)).

As a replacement for knob-and-tube wiring, nonmetallic-sheathed cable was introduced in 1928. This wiring method had an outer covering which provided additional protection for the conductors (Article 334).

The purveyors of the earliest forms of overcurrent protection were met with a dilemma in designing the best methods of protection. In this respect, it is important to consider the melting temperatures of various metals. The following is a list of melting and fusing temperatures of different materials.

Fahrenheit	Celsius
Aluminum	
1220	660
Silver	
1762	961
Bronze	
1562	850
Zinc	
786	419
Brass	
1652	900
Steel	
2498	1370
Gold	
1945	1063
Platinum	
3218	1770
Copper	
1981	1083
Solder (electrical/electronic work)	
370	188

From this list, it is easy to see why solder is used in the manufacture of fuses. It has a low melting temperature as compared to copper or aluminum and this makes it a logical choice. The electrical conductivity of this material is very good where properly applied. Other materials include zinc, copper, silver, aluminum, or alloys which add to the stability of the fuse.

Based on the definition of the term overload, we can readily see that an overload may be sustained for long periods, which create higher levels of heat in conductors and equipment. And depending on the operating characteristics of the fuse or circuit breaker, the device may not clear. Or, it may take a disproportionate amount of time to open. We know that thermal stress in conductor insulation will reduce its service life. As an example, thermoplastic insulation with a 75 degree C. temperature limit may have an average service life of 30 years. However if this insulation is subjected to a sustained temperature of 5-10 degrees C. above 75 degrees C., the service life of the insulation may be reduced by 50%.

A typical squirrel-cage induction motor has a service life of 20 years. However, if a motor with a marked temperature rise of 40 degrees C. is operated at 5 degrees C. above 40 degree C., the service life is reduced by 50%. And for each 5 degrees C. temperature increase, the service life is reduced by an additional 50%.

Obviously, this level of thermal stress cannot be tolerated. The solution is to design the electrical components in such a way that damage will not be a problem. This includes normal operating conditions, as well as the consideration of the effects of the continuous loading of conductors and equipment.

But what about abnormal loading conditions? How does the definition of the term overcurrent factor into the protection of conductors and equipment? How can we determine the withstand rating of conductor insulation for short time periods?

The answer is relatively easy to determine with the understanding of some basic concepts.

First, there must be a comprehensive study of a particular electrical system to determine the short-circuit current that is available at different locations.

Next, we must determine the operating characteristics of the circuit components. This includes the overcurrent devices, wire, and equipment.

The following Table identifies conductor insulation short-time withstand ratings. This is based on the use of copper conductors with 75 degree C. insulation, and a short-circuit temperature of 150 degrees C. The time constants range from 5 seconds to 1/8 cycle (.002 seconds). And the wire sizes range from 14 AWG to (2.08mm^2) to 500 kcmil (253mm^2) copper.

Also, there are tables relating to equipment terminal limits at 250°C. and conductor fusing currents at 1083°C.

Insulation Withstand Ratings 150° C. Maximum

AWG	Normal 75°C.	5 Second 150° C.	1 Second 150° C.	1 Cycle– .016 Sec. 150° C.	1/2 Cycle – .008 Sec. 150°C.	1/4 Cycle – .004 Sec. 150° C.	1/8 Cycle – .002 Sec. 150° C.
14	20A	97A	217A	1,715A	2,425A	3,429A	4,850A
12	25A	155A	347A	2,740A	3,875A	5,480A	7,750A
10	35A	246A	550A	4,349A	6,150 A	8,697A	12,300A
8	50A	397A	888A	6,912A	9,775A	13,824A	19,550A
6	65A	621A	1,389A	10,978A	15,525A	21,956A	31,050A
4	85A	988A	2,209A	17,466A	24,700A	34,931A	49,400A
3	100A	1,245A	2,784A	22,008A	31,125A	44,017A	63,450A
2	115A	1,571A	3,513A	27,772A	39,275A	55,543A	78,550A
1	130A	1,981A	4,430A	35,019A	49,525A	70,039A	99,050A
1/0	150A	2,499A	5,588A	44,176A	62,475A	88,353A	124,950A
2/0	175A	3,150A	7,044A	55,685A	78,750A	111,369A	157,500A
3/0	200A	3,972A	8,882A	70,216A	99,300A	140,431A	198,600A
4/0	230A	5,008A	11,198A	88,530A	125,200A	177,060A	250,400A
250 kcmi	25SA	5,917A	13,231A	104,599A	147,925A	209,198A	295,850A
300 kcmi	285A	7,101A	15,878A	125,529A	177,525A	251,058A	355,050A
350 kcmi	310A	8,284A	18,524A	146,442A	207,100A	292,884A	414,200A
400 kcmi	335A	9,467A	21,169A	167,354A	236,675A	334,709A	473,350A
500 kcmi	380A	11,834A	26,462A	209,198A	295,850A	418,395A	591,700A

Example

Insulation Withstand Rating

No. 6 AWG copper – 65 amperes (continuous) – 75°C
No. 6 AWG copper – 26,240 circular mils – 13.30mm^2
I^2T – ampere – squared seconds
I^2T – one ampere for every 42.25 circular mils of conductor cross-sectional area for 5 seconds

$$No.6\ AWG - \frac{26{,}240\, circular\ mils}{42.25} = 621\, amperes - 5\, seconds$$

To determine the insulation withstand rating for 1 cycle (.016 seconds), the calculation is as follows:
621 amperes × 621 amperes x 5 seconds = 1,928,205

$$\frac{1{,}928{,}205}{.016} = 120{,}512{,}813$$

$$\sqrt{120{,}512{,}813} = 10{,}978\, amperes$$

Therefore, the one-cycle insulation withstand rating is 10,978 amperes, which will produce a temperature of 150°C. This is the maximum temperature that the insulation can safely withstand without damage.

Terminal Withstand Ratings 250° C. Maximum

AWG	Normal 75°C.	5 Second 250° C.	1 Second 250° C.	1 Cycle – .016 sec. 250° C.	1/2 Cycle – .008 sec. 250° C.	1/4 Cycle – .004 sec. 250° C	1/8 Cycle– .002 sec. 250° C.
14	20A	112A	250A	1,980A	2,800A	3,960A	5,600A
12	25A	178A	398A	3,147A	4,450A	6,293A	8,900A
10	35A	282A	631A	4,985A	7,050A	9,970A	14,100A
S	50A	449A	1,004A	7,937A	11,225A	15,875A	22,450A
6	65A	714A	1,597A	12,622A	17,850A	25,244A	35,700A
4	85A	1,136A	2,540A	20,082A	28,400A	40,164A	56,800A
3	100A	1,459A	3,262A	25,792A	36,475A	51,583A	72,950A
2	115A	1,806A	4,038A	31,926A	45,150A	63,852A	90,300A
1	130A	2,277A	5,092A	40,252A	56,925A	80,504A	113,850A
1/0	150A	2,873A	6,424A	50,788A	71,825A	101,576A	143,650A
2/0	175A	3,622A	8,099A	64,029A	90,550A	128,057A	181,100A
3/0	200A	4,566A	10,210A	80,716A	114,150A	161,432A	228,300A
4/0	230A	5,758A	12,875A	101,788A	143,950A	203,576A	287,900A
250kcmil	255A	6,803A	15,212A	120,261A	170,075A	240,522A	340,150A
300 kcmil	285A	8,163A	18,253A	144,303A	204,075A	288,606A	408,150A
350 kcmi	310A	9,524A	21,296A	168,362A	238,100A	336,724A	476,200A
400 kcmi	335A	10,884A	24,337A	192,404A	272,100A	384,808A	544,200A
500 kcmi	380A	13,605A	30,422A	240,505A	340,125A	481,009A	680,250A

Example

Terminal Withstand Rating

No. 6 AWG copper – 65 amperes (continuous) – 75°C
No. 6 AWG copper – 26,240 circular mils – 13.30mm^2
I^2T – ampere-squared seconds
I^2T – one ampere for every 36.75 circular mils of conductor cross-sectional area for 5 seconds

$$No.6\ AWG - \frac{26,240\,circular\ mils}{36.75} = 714\,amperes - 5\,seconds$$

To determine the conductor terminal withstand rating for one cycle (.016 seconds), the calculation is as follows:

714 amperes × 714 amperes × 5 seconds = 2,548,980

$$\frac{2,548,980}{.016} = 159,311,250$$

$$\sqrt{159,311,250} = 12,622\,amperes$$

Therefore, the one-cycle (.016 seconds) terminal withstand rating is 12,622 amperes, which will produce a temperature of 250°C. This is the maximum temperature that the terminal can safely withstand.

Fusing or Melting Current 1083° C. Maximum

AWG	Normal 75° C.	5 Second 1083° C.	1 Second 1083° C.	1 Cycle – .016 Sec. 1083° C.	1/2 Cycle – .008 Sec. 1083° C.	1/4 Cycle– .004 Sec. 1083° C.	1/8 Cycle – .002 Sec. 1083° C.
14	20A	254A	568A	4,490A	6,350A	8,980A	12,700A
12	25A	403A	901A	7,124A	10,075A	14,248A	20,150A
10	35A	641A	1,433A	11,331A	16,025A	22,663A	32,050A
8	50A	1,020A	2,281A	18,031A	25,500A	36,062A	51,000A
6	65A	1,621A	3,625A	28,656A	40,525A	57,311A	81,050A
4	85A	2,578A	5,765A	45,573A	64,450A	91,146A	128,900A
3	100A	3,312A	7,406A	58,548A	82,800A	117,097A	165,000A
2	115A	4,101 A	9,170A	72,461A	102,475A	144,922A	204,950A
1	130A	5,169A	11,558A	91,376A	129,225A	183,105A	258,950A
1/0	150A	6,523A	14,586A	115,311A	163,075A	230,623A	326,150A
2/0	175A	8,221A	18,383A	145,328A	205,525A	290,656A	411,050A
3/0	200A	10,364A	23,175A	183,211A	259,100A	366,423A	518,200A
4/0	230A	13,070A	29,225A	231,047A	326,750A	462,094A	653,500A
250 Kcmil	255A	15,442A	34,529A	272,979A	386,050A	545,957A	772,100A
300 Kcmi	285A	18,530A	41,434A	327,567A	463,250A	655,134A	925,500A
350 Kcmi	310A	21,618A	48,339A	382,156A	540,450A	764,312A	1,080,900A
400 Kcmi	335A	24,707A	55,247A	436,762A	617,675A	873,524A	1,235,350A
500 Kcmil	380A	30,883A	69,056A	545,939A	772,075A	1,091,879A	1,544,150A

Example

Fusing or Melting Current

No. 6 AWG copper – 65 amperes (continuous) – 75°C
No. 6 AWG copper – 26,240 circular mils – 13.30mm^2
I^2T –ampere –squared seconds
I^2T – one ampere for every 16.19 circular mils of conductor cross-sectional area for 5 seconds

$$No.6\ AWG - \frac{26,240\,circular\,mils}{16.19} = 1621\,amperes - 5\,seconds$$

To calculate the fusing or melting current for one cycle (.016 seconds), the calculation is as follows:

1621 amperes × 1621 amperes × 5 seconds = 13,138,205

$$\frac{13,138,205}{.016} = 821,137,812$$

$$\sqrt{821,137,812} = 28,656\,amperes$$

Therefore, the one-cycle fusing or melting current of this conductor is 28,656 amperes, which will produce a temperature of 1083°C. Based on these conditions, the conductor will fuse or melt in 1 cycle.

AWG - To Metric Conversion Chart

AWG	Circular Mil Area	Metric Size – MM^2
14	4110 cm	2.08
12	6530 cm	3.31
10	10,300 cm	5.261
8	16,510 cm	8.367
6	26,240 cm	13.3
4	41,740 cm	21.15
3	53,620 cm	26.67
2	66,360 cm	33.62
1	83,690 cm	42.41
1/0	105,600 cm	53.49
2/0	133,100 cm	67.43
3/0	167,800 cm	85.01
4/0	211,600 cm	107.2
250 kcmil	250,000 cm	127
300 kcmil	300,000 cm	152
350 kcmil	350,000 cm	177
400 kcmil	400,000 cm	203
500 kcmil	500,000 cm	253

$$\text{e.g.} - \text{No.2 AWG} = \frac{66,360\ \text{Circular Mils}}{1973.53} = 33.625\text{mm}^2$$

For example, a 4 AWG (21.15mm^2) copper conductor has a continuous current rating of 85 amperes at 75 degrees C. (Table 310.15(B)(16)). But, for 5 seconds, it has a rating of 988 amperes, with no insulation damage (150 degrees C.). And, for a period of one cycle (.016 seconds) the ampere rating is 17,466 amperes. As a matter of reference, for 50Hz systems, one cycle equals .020 seconds. So, insulation withstand ratings at 50Hz will be lower for the same time period.

After calculating the available fault-current and determining the operating characteristics of overcurrent devices, we can use this information to select the proper conductor size for different applications.

In recent NEC cycles there have been new references that relate to the Short-Circuit Current Rating of equipment and the Available Fault Current. These references are as follows:

Section 110.24(A) - Available Fault-Current at Service Equipment (in other than dwelling occupancies)
Section 230.82(3) - Meter Disconnect Switches
Section 285.6 - Surge-Protective Devices
Section 409.22 - Industrial Control Panels (UL-508)
Section 430.8 - Motor Controllers
Section 430.130(A)-Power Conversion Equipment
for Adjustable–Speed Drive Systems
Section 440.4(B) - HVAC-Multimotor and Combination Loads
Section 670.3(A) - Industrial Machinery Electrical Panel (NFPA 79)
Section 700.5(E) – Transfer Equipment
Section 701.5(D) – Transfer Equipment

These references require that the available fault-current be marked on the equipment. So, the available fault-current must be determined and this information must be identified. And, if alteration are made to the supply system, such as the replacement of a transformer with a different percent impedance, the available fault-current must be recalculated and the equipment must be marked with this new information.

A very important, but often misunderstood NEC reference is Section 240.4, which covers Protection of Conductors. And it states that conductors, other than flexible cords, flexible cables (400.4), and fixture wires (402.5) shall be protected at the ampacities specified in 310.15, unless otherwise permitted in 240.4(A) through (G).

A commonly used reference is 240.4(B), where a conductor(s), other than a branch-circuit supplying more than one receptacle for cord-and-plug connected portable loads, has an ampere rating that does not correspond to the standard ampere rating of a fuse or circuit breaker, the next standard size of overcurrent device may be used. This permission does not apply above 800 amperes.

For example, a 6 AWG (13.30mm^2) copper THW insulated conductor, which has a normal ampacity of 65 amperes (Table 310.15(B)(16), protected by a 70 ampere overcurrent device (240.6(A)).

The point is that it is not acceptable to simply calculate an electrical load and select the conductor size from Table 310.15(B)(16). Unfortunately, this is exactly how this process is normally done. The conductor ampacities listed in this Table (and the other Tables of Article 310) are based on the conditions expressed in these tables. However, the Informational Note of 240.4 references ICEA P32-382-2007 for information on the allowable short-circuit currents for insulated copper and aluminum conductors. I am proud to say that I submitted the Proposal which led to this Informational Note during the 2011 NEC cycle. While Informational Notes are not enforceable requirements of the NEC (90.5(C)), the concept of providing protection from short-circuits and ground-faults is addressed in Section 110.10-Circuit Impedance, Short-Circuit Ratings, and Other Characteristics. We addressed this issue for insulated copper conductors at the beginning of this Chapter. In addition, we will calculate the available fault-current at different locations in a typical distribution system in a later chapter.

Section 240.1-Scope-Parts I through IV cover the requirements for overcurrent protection where the voltage is up to 1000 volts. Part VIII covers overcurrent protection for Supervised Industrial Installations where the voltage does not exceed 1000 volts.

Part IX addresses overcurrent protection where the voltage exceeds 1000 volts.

Once again, the concept of providing proper overcurrent protection is to disconnect a circuit when the current exceeds the rating of an overcurrent device (time vs. current) in order to eliminate overheating and damage to the circuit components.

But it is also important to determine the interrupting rating (110.9) and fault let-through current of the overcurrent device (110.10).

Overcurrent Protection-Over 1000 Volts

240.100 (A) - Feeders and branch-circuits are required to have overcurrent protection in each ungrounded conductor. And, much the same as in 240.21, the overcurrent device is to be located where these conductors receive their supply. However, if the system design is under engineering supervision that is based on the consideration of fault studies and time-current coordination analysis of the protective devices and the associated conductor damage curves, an alternative location in the circuit is acceptable.

There are three classifications of fuses referenced in ANSI C37.40. These include the following:

1. General Purpose Current Limiting – This fuse is capable of interrupting all currents from the rated interrupting current of the fuse, down to the current that causes the melting of the fuse element in one hour.
2. Back-up Current Limiting Fuse – This fuse is able to interrupt all currents from the maximum rated interrupting current down to the rated minimum interrupting current.
3. Expulsion Fuse – This is a vented fuse that is designed to expel gases produced by the arc and lining of the fuse holder, which may be aided by a spring in the opening process.

There are additional rules that apply to Expulsion Fuses. For example, the total clearing curve of a downstream protective device must be below a curve that represents a value of 75% of the minimum melting curve of the Expulsion Fuse.

And, the same applies to any upstream protective device so that the Expulsion Fuse total clearing curve is below 75% of the minimum melting curve of the upstream protective device.

For current-limiting fuses, the consideration is that the arc voltage during interruption does not exceed the insulation voltage rating of the system. Usually, if the fuse voltage rating does not exceed 140% of the system voltage, there should not be a problem.

Current-limiting overcurrent devices at lower voltages also produce a voltage-rise as they interrupt fault currents. But, the arc voltage is restricted to 6000 volts.

240.101 (A) permits the continuous ampere rating of a fuse to be 3 times the conductor ampacity. The minimum trip setting of an electronically actuated fuse or the long-time trip element setting of a circuit breaker is not to exceed 6 times the conductor ampacity.

240.101(B) recognizes that feeder tap conductors are permitted to be protected by the feeder overcurrent device, where this device also protects the tap conductor.

The same rules for Expulsion Fuses should be applied to current-limiting fuses, that is, coordination with downstream and upstream protective devices. And, where other current-limiting fuses are installed in the system, the minimum I^2T (current squared times time) melting of the fuse should be greater than the total clearing I^2T of the downstream current-limiting fuse.

3

Motor Circuits

The protection of motor circuits requires a careful analysis of the various types of overcurrent devices that are used for this application.

Due to the initial starting current of a motor, which may be six or seven times (or more) of the motor normal running current, time-delay devices lend themselves very effectively to motor circuits. For example, a time-delay fuse is designed to carry 5 times its rating for at least 10 seconds. This type of device is perfect for most motor installations. And, because of this time-delay feature, the fuse rating may be low enough to afford back-up protection for the motor thermal overload devices, or these devices may be used in lieu of the thermal overload devices (430.32).

Single-phasing has always been a problem for 3-phase motors. The loss of one-phase may cause the current in the other phases to increase to 173% (1.732), or even 200% of the rated full-load current of the motor. If time-delay fuses are used as the motor overcurrent protection, and they are sized at 115% or 125% of the motor full-load current rating, they may serve as a means of protection against overload damage.

In addition, the use of smaller time-delay fuses would permit the use of lower amperage rated switches, which may cost only one-third that of the switches required for large single-element fuses. And the smaller fuses would also cost less than the larger single-element devices.

Due to the fact that the lower ampere rated time-delay fuses are more closely matched to the ampere rating of the connected load, they also provide better short-circuit protection than larger ampere rated single-element fuses because the fault-current must increase to higher levels before the larger fuse reacts to interrupt the fault.

And selective coordination is easier to apply with time-delay fuses than with non-time-delay fuses because the fuse ratings may be a closer match between upstream and downstream fuse ratings.

Generally, Table 430.52 permits a time-delay fuse to have a maximum rating of 175% for single-phase and polyphase motors. And for rare cases where 175% may not be sufficient for the motor starting and running conditions, the fuse rating may be increased to 225% (430.52(C)(1), Exception No. 2(b)). But, for typical motor applications, these ratings will be much lower for better overcurrent protection.

And finally, where motors are installed in areas where there are higher ambient temperatures, the operating characteristics of a time-delay fuse will afford better protection and require less derating than non-time delay devices.

For transformer applications, the primary inrush-current may be 20 times the normal full-load primary current. Even though this condition will exist for only a very brief period, properly sized time-delay fuses will not be affected by this inrush current.

For constant voltage DC motors, the time-delay fuse may be rated at 150% of the motor full-load current rating (Table 430.52).

Inverse-time circuit breakers are also effective for motor circuits. As their name implies, the clearing time of the circuit breaker decreases as the motor circuit current increases. These circuit breakers may be rated at up to 250% of the motor full-load current. And for sustained periods of starting, where 250% is not sufficient, this rating may be increased to 400% for full-load currents of up to 100 amperes, and 300% for full-load currents of over 100 amperes (430.52(C)(1), Exception No. 2(c)).

It should be noted that all overcurrent devices have an inverse-time characteristic. In the case of inverse-time circuit breakers, there is an overload or thermal unit and an instantaneous-trip function. The overload function typically operates through the use of a bimetallic element which is made of differing alloys. And these metallic elements react to the heat associated with an overload, with one element bending to the extent that it causes the release of the tripping mechanism. In addition to the thermal element, these devices have a magnetic element which recognizes the high currents associated with a fault-current. For installations where the fault-current is limited, the pick-up setting is normally at 7 times the rating of the circuit breaker. For installations where the fault-current is expected to be much higher, such as in industrial or institutional installations, the pick-up setting may be 12 times the rating of the circuit breaker.

Of course, it would be an advantage to keep the rating of the circuit breaker as close as possible to the actual connected load without the problem of nuisance tripping. And it would be an extreme condition to need an inverse-time circuit breaker rated at 300% or 400% of the motor full-load current rating.

Nontime delay (single-element) fuses may be rated at up to 300% of the motor full-load current rating. But, where this is not sufficient for the motor circuit conditions, the maximum fuse rating may be increased to 400% of the motor full-load current rating (430.52(C)(1), Exception No. 2(a)).

It should be noted that Class CC time-delay fuses are permitted to have the same rating as nontime delay fuses. The operating characteristics of Class CC fuses are relative to a very rapid clearing time. So, higher ratings may be necessary for certain applications, which include the starting currents of motors.

Instantaneous-Trip Circuit Breakers are, as the name implies, designed to trip without any time-delay. However, where motor inrush-current, due to short current transients, may cause the inadvertent tripping of the circuit breaker, a damping means may be used. The settings of these devices are permitted to be very high (by Exception). For Design B energy-efficient motors, as high as 1700% of the motor full-load current rating. For this reason, their use is restricted to motor controllers, with a contactor and proper overload protection (430.52(C)(3)). The reasoning for this is that this circuit breaker is not designed to recognize overload conditions, but only short-circuit or ground-fault conditions. These circuit breakers are sometimes referred to as Motor Circuit Protectors. They are adjustable to accommodate the motor starting current, and the interrupting rating of the assembly will be identified on the cover of the unit.

Keep in mind that the overcurrent device that is used to provide protection for motors also must be designed to provide short-circuit and ground-fault protection for all of the circuit components, including wire, switches, controllers, and the overload protective devices. Adjusting the trip-setting of the instantaneous-trip circuit breaker at 8-10-12, or more, of the motor full-load current rating may be questionable with respect to providing the required overcurrent protection for the circuit components.

And now for the requirements that apply to conductors that are used to supply single continuous-duty motors. Section 430.22 requires that conductors supplying a single motor must have an ampacity of 125% of the motor full-load current rating. This ampere rating is not taken from the motor nameplate, but from the appropriate table of Article 430. The table values will normally be above the nameplate current rating. The nameplate ampere rating is only used to size the overload devices. And even though the motor starting current may be quite high, the conductor insulation short-circuit current rating should be sufficient to prevent damage. However, this must be determined by calculation, along with the other circuit components.

Where more than one motor is supplied, the provisions of 430.24 apply. In addition, this section applies where motors and nonmotor loads are supplied by the branch-circuit. In this case, the supply conductors are sized at 125% of the ampere rating of the largest motor, plus the sum of the other motor or other loads. If two or more motors of the group have the same ampere rating, only one is used as the largest motor, and the rest of the load is added accordingly.

Adjustable-Speed Drive Systems

The protection of adjustable-speed drive systems from UL 508C, involves a standard fault current test where the test currents are 5000 amperes for 1.5-50 hp drives and 10,000 amperes for 51-200 hp drives. The drive does not have to be operational after the fault current test, and the maximum short-circuit current rating must be marked on the equipment. If a specific type of overcurrent device is specified on the equipment, only that type of device may be used.

There is a high current fault test that is administered following the standard fault current test. This test can be at any level of fault current above the standard fault current test. Once again, the device may not be operational after the test.

The best level of protection for this type of equipment is Type 2 protection. In this case, the drive is tested and marked with a high short-circuit current rating. There can be no damage to the equipment and it must be completely operational after the fault has been cleared by the overcurrent device.

On this type of circuit, the ampere rating of the disconnecting means must be no less than 115% of the drive unit Rated Input Current (430.128). The drive unit supply conductors are sized at no less than 125% of the Rated Input Current (430.122(A)). It is typical that the type of protective device, along with its rating and setting are identified on the equipment in accordance with 430.130(A)(1)(2)(3).

As we already know, an overcurrent exists when the normal load current is exceeded. The cause of the overcurrent condition may be in the form of a short-circuit, ground-fault, or an overload. A short-circuit or ground-fault will cause the current to flow through a path of the least resistance or impedance, typically by-passing the connected load on its path back to the source. Of course, depending on the type of the circuit protection and the speed of response to the overcurrent, as well as other characteristics, such as current limitation, the circuit components may, or may not, be damaged.

For motor circuits, another important factor is unbalanced voltages on 3-phase equipment. For example, a voltage unbalance of 5% may produce a temperature rise of up to 60°C. in the phase with the highest current, which if not promptly cleared by the overload device, serious damage may be the result. This problem may be for worse if the motor overload protection is provided by a fuse or circuit breaker and not by separate overload devices. If the fuse is a non-time-delay type or a thermal magnetic circuit breaker, sized at 115% or 125% of the motor nameplate current rating, in accordance with 430.32(A)(1), these overcurrent devices may, or most likely, will trip during the motor starting (inrush) current. Even inverse-time circuit breakers, set at 115% or 125% of the motor nameplate current rating may be affected by the motor inrush current. Time-delay fuses are designed to carry current at five times their rating for at least ten seconds. So, these devices should be capable of carrying inrush currents without opening. This is an important consideration, as motor starting or inrush currents may exceed 20 times the normal full-load current during the first ½ cycle (0.008 seconds) of operation, and then subside to 6-8 times of the full-load current for possibly several seconds.

Another important consideration is the possibility of the loss of one-phase (single-phasing) on a delta supply system. Where this condition occurs on the secondary side of the transformer and the motor(s) is running, the affected lost phase will carry 0 amperes, while the other phases will be subjected to a current of 173% of the normal full-load current. Under this running condition, the motor attempts to drive the connected load, resulting in either significant damage to the motor, or until the overload elements or the motor overcurrent device operates, and the circuit is open.

In some instances, the current in the two remaining phases may be much higher than the theoretical 173% of the normal full load-current. The power factor may change, especially if the connected motor load is quite high. This may cause the current in the two phases to increase to as much as 200%, and in some cases, the current may reach locked-rotor conditions.

The NEC permits the motor overload protection to be based on the motor nameplate current rating. However, even better protection is provided when the overload protection is based on the actual running current of the motor.

Earlier, we mentioned that a voltage unbalance will produce an increase in operating temperature. For a three-phase motor, the temperate-rise can be calculated. Recently, in checking the voltages on a three-phase, 480 volt system, the voltage measured:

Phase A-B-476 volts
Phase B-C-471 volts
Phase A-C-468 volts

$$476 + 471 + 468 = 1415 \text{ volts}$$

$$\frac{1415}{3} = 472 \text{ Volts(average voltage)}$$

$$476 \text{ (highest voltage)} - 472 \text{ (average voltage)} = 4\text{V}.$$

$$\frac{4\,volts\ (Voltage\,difference)}{472(average\,voltage)} = 0.008 \times 100 = .85\%$$

$2x(.85)^2 = 1.445$ Percent Temperature Rise. This is expected temperature rise in the phase winding with the highest current (Phase A-C). For a motor with a temperature rise of 40°C., the unbalanced voltage will result in a temperature rise in this winding of:

$$\begin{array}{r} 40°\text{C.} \\ \times 1.445 \\ \hline 57.8°\text{C.} \end{array}$$

4

Selective Coordination

Beginning with the 1993 NEC cycle, selective coordination for certain elevator installations became mandatory. Section 620.62 states that 'where more than one driving machine disconnecting means is supplied by a single feeder, the overcurrent protective devices in each disconnecting means shall be selectively coordinated with any other supply side overcurrent protective devices'.

The selective coordination is required to be selected by a licensed professional engineer or other qualified person engaged primarily in the design, installation, or maintenance of electrical systems.

The selection must be documented and made available to those authorized to design, install, inspect, maintain, and operate the system.

Since the 1993 cycle, selective coordination has been expanded to include the following:

645.27 – Information Technology Equipment. This applies to Critical Operations Data Systems. These systems require continuous operation for reasons of public safety, emergency management, national security, or business continuity (645.2).

695.3 – Power Source(s) for Electric-Motor Driven Fire Pumps – Where multiple overcurrent devices are in series with the fire pump supply, 695.3(C)(3) requires that these overcurrent devices be selectively coordinated with the supply side overcurrent devices.

700.32 – Emergency System Overcurrent Devices shall be selectively coordinated with all supply-side overcurrent devices. Because of the critical nature of the emergency system, that is, life safety, Article 700 applies in conjunction with NFPA 99-2015-Health Care Facilities Code, NFPA 101-2015- Life Safety Code, NFPA 110-2013- Standard for Emergency Standby Power Systems, and ANSI/UL 1008-Transfer Switch Equipment. Selective coordination is extremely important here, as the loss of even a part of this system may result in death, serious injury, or catastrophic loss of equipment.

701-27 – Legally Required Standby Systems, that is, systems required by municipal, state, federal, or other codes, or by any governmental agency having jurisdiction.

The Informational Note of 701.2 identifies examples of the types of load supplied by this system, including communications systems, heating and refrigeration systems, and ventilation and smoke removal systems. The loss of these systems may affect rescue and/or fire-fighting operations.

708.54 – Critical Operations Power Systems - The overcurrent devices for these systems must be selectively coordinated with all supply-side overcurrent protective devices.

On another note, where ground-fault protection is required for the operation of the service and feeder disconnecting means, the feeder GFP must be fully selective with the service GFP so that the downstream device operates to clear a ground-fault on its load side before affecting the service GFP device. This selectivity applies to Health Care Facilities (517.17(C)) and Critical Operations Power Systems (708.52(D)).

Where selective coordination is applied for fuses, the concept appears to be relatively simple, that is, the total clearing energy of a downstream fuse is less than the melting energy of the upstream fuse. The available short-circuit current must be calculated. And the thermal energy is directly proportional to the square of the current multiplied by the I^2T (amperes squared x time) clearing time of the overcurrent device. For example, a 5000 ampere fault, which is cleared in ½ cycle (.008 seconds), would produce a melting energy 200,000 ampere-squared seconds (5000 amperes × 5000 amperes × .008 seconds = 200,000). This is the melting energy of the fuse, and the total clearing energy is equal to the melting time, plus the arcing time. So, the clearing energy (I^2T) of the downstream fuse must be less than the melting energy (I^2T) of the upstream fuse.

This seems to be a tedious process, and at one time, where a designer would compare the time vs. current curves of in-line fuses on a light table, it certainly was.

However, one fuse manufacturer (Bussmann) has produced a fuse selectively ratio guide that has greatly simplified this process.

The selectivity ratios apply where the available fault current is less than 200,000 amperes, or the interrupting rating of the fuses, whichever is less. For example, a line-side Bussmann dual element low-peak 400 ampere LPN-RK SP fuse will be selectively coordinated with a load-side Bussmann dual element low peak LPJ-SP TCF fuse where a ratio of 2:1, or greater is maintained. If the downstream fuse has a rating of 200 amperes, or lower, these fuses are selectively coordinated. That is, the downstream fuse element will melt, arc, and totally clear, before the upstream fuse element begins to melt.

Using the fuse selectively ratios, the designer or installer just needs to size the fuses properly, based on the connected load, and the available short-circuit current.

Circuit Breakers

There is a simplified approach to identify circuit breaker coordination where the breakers are of the instantaneous-trip type. By multiplying the instantaneous-trip setting times the circuit breaker ampere rating, the result will be the approximate point where the circuit breaker enters the instantaneous-trip region. This approach is applicable to the instantaneous-trip function, and not to the overload region. In most cases, this method will cover the overload region anyway.

As an example, where a 600 ampere circuit breaker has an instantaneous-trip set at 10 times its ampere rating, or 6000 amperes, this device will unlatch and open. If this circuit breaker is on the line-side of a 400 ampere circuit breaker that has its instantaneous-trip set at 10 times its ampere rating, or 4000 amperes, these 2 devices are selectively coordinated, as long as the available short-circuit current does not equal 6000 amperes, where both circuit breakers will open.

This procedure is acceptable, but there is another consideration which relates to a tolerance region. This is a + or – region, which may range from + or – 20% for a thermal magnetic circuit breaker with high-trip setting, + or – 25% for a thermal magnetic circuit breaker with a low-trip setting. Or a + or – tolerance of 10% for a circuit breaker with an electronic trip. However this information should be based on information from the manufacturer. The positive tolerance may be ignored and only the negative tolerance is considered when using this simple method.

For example, consider an electronic-trip circuit breaker with a rating of 1000 amperes with an instantaneous-trip set at 10 times and a tolerance of 10 %, and a downstream 400 ampere thermal magnetic circuit breaker set at 10 times and a tolerance of 25%.

$$\left(400\,amperes \times 10\right) \times \frac{(1-25\%)}{100}$$

$$(4000) \times 1 - 0.25) = 4000\ amperes \times .75 = 3000\ amperes$$

For an overcurrent of less than 3000 amperes, the 400 ampere circuit breaker will coordinate with downstream circuit breakers.

The 1000 ampere circuit breaker with the instantaneous-trip set at 10 times its rating and a tolerance of 10%, the coordination will be:

$$\left(1000 \times 10\right) \times \frac{(1 - 10\%)}{100}$$

$$(10{,}000) \times (1 - 0.10) = 10{,}000 \times 0.9 = 9000 \textit{ amperes}$$

For an overcurrent of less than 9000 amperes, the 1000 ampere circuit breaker will coordinate with the downstream 400 ampere circuit breaker. But for short-circuit currents of 9000 amperes or more, both circuit breakers will open.

Taking these tolerances into consideration, the selective coordination is more accurate.

Circuit breaker manufacturers publish circuit breaker to circuit breaker coordination tables based on testing. Of course, a short-circuit analysis must be done to calculate the available short-circuit current. Then, based on this calculation, it may be determined if the circuit breakers will coordinate.

Electronic-trip and insulated-case circuit breakers may have a short-time delay function which allows the circuit breaker a period of time, normally from 6-30 cycles (0.096 – 0.480 seconds), to clear a fault current. This feature is beneficial for low-level fault currents.

However, for higher levels of fault current, there is an instantaneous-trip function, set by the manufacturer at 8-12 times the circuit breaker rating. This instantaneous-trip feature must be considered when determining the selective coordination of the entire system.

Beginning with the 1971 NEC cycle, ground-fault protection of equipment became a part of Article 230, specifically in 230.95. This protection involved solidly grounded, 3 phase, 4 wire, wye connected systems, operating at up to 600 volts, phase-to-phase, and, over 150 volts, phase-to-neutral, where the rating of the service disconnect is 1000 amperes, or more. Now this protection applies to voltages of more than 150 volts-to-ground, but not exceeding 1000 volts phase-to-phase for each service disconnect rated 1000 amperes, or more. Since 1971, this type of protection has been expanded to include feeders (215.10), the building or structure main disconnecting means (240.13), 517.17 for Health Care Facilities, and 708.52 for Critical Operations Power Systems, where two levels of ground-fault protection are required, and 700.6(D) for Emergency Systems, as well as 701.6(D) for Legally-Required Standby Systems. However, for these last two systems, there will be a ground-fault sensor, located at, or ahead of the main disconnecting means. And the maximum setting of the signal devices shall be for a ground-fault current of 1200 amperes (700.6(D)). This same provision applies to the legally-required standby source (701.6(D)). In these cases, it is not required to have automatic ground-fault protection, but a means of identifying a ground-fault condition, so that it may be cleared when the normal service is restored (700.31 and 701.26).

Of course, where ground-fault protection is required, selective coordination must be considered. In fact, it may be better to assure the selective coordination of the downstream overcurrent devices, before considering the settings of the GFP. These settings include the pick-up setting associated with the ampere setting and the time delay setting. This equipment only recognizes ground-faults and provides no protection against short-circuits (phase-to-phase and phase-to-neutral).

The current limit referenced in 230.95 is 1200 amperes, with a maximum time-delay of one second for ground-faults of 3000 amperes, or higher. The ground-fault protective relay trip-setting must be selectively coordinated with the downstream overcurrent devices to eliminate the possibility of a total system 'blackout'. Unfortunately, when this has happened, sometimes due to the fact that the ampere setting, as well as the instantaneous-trip setting, are too low, this protection scheme is compromised and the ground-fault protection is lost.

If it is a problem to selectively coordinate the ground-fault protection with the downstream overcurrent devices, it may necessitate the use of multiple smaller overcurrent devices in lieu of one large feeder or service overcurrent device. For example, 5-800 ampere overcurrent devices, as opposed to 1-4000 ampere overcurrent device, or, the use of an ungrounded or impedance-grounded system, with a solidly-grounded system located downstream from the source.

For Health Care Facilities and Critical Operations Power Systems there must be two levels of ground-fault protection (517.17 and 708.52). And these two levels of protection must be fully selective. At one time, 517.17 specified at least a 6-cycle separation between the upstream ground-fault protection relay time band and the downstream ground-fault relay. Now, 517.17(D) and 708.52(D) require that the downstream (feeder) ground-fault protection opens on ground faults on the load side of the feeder device without affecting the ground-fault protection at the service. Separation of the ground-fault time-current characteristics shall be in conformance with the manufacturer's recommendations to achieve 100% selectivity.

In any event, the ground-fault protection must be tested when first installed to ensure that each level of protection is operational (517.17(D)(708.52(C)).

Designing the distribution system to be fully selective, including where ground-fault protection is required, can be quite difficult. This is especially true where ground-fault protection of equipment is required. Short-circuit current calculations must identify the available short-circuit currents throughout the distribution system. This will identify the required interrupting ratings of overcurrent devices (110.9), as well as the required operating characteristics of the overcurrent devices installed at various locations, even at the farthest points from the source.

However, in considering ground-fault conditions, it must be recognized that, by far, the vast majority of ground-faults are arcing type faults. That is, a strike and restrike between an ungrounded conductor and the equipment grounding system. This may be caused by damage to conductor insulation, or by contaminants absorbed through the insulation, possibly dirt combined with moisture that form tracking paths through the insulation, and possibly, to the equipment grounding system. There is a voltage-drop in this arcing fault, possibly 50 volts, or more, and this compounds the problem in returning the ground-fault current to the source. If not promptly cleared in a system that is solidly grounded, the arcing fault may lead to equipment damage, or possibly result in a short-circuit, which may cause more severe damage.

In an ungrounded system, this fault, which is recognized and identified by the ground detectors required by 250.21(B)(1)(2), must be located and corrected before the arcing ground-fault causes damage to an opposite phase, resulting in more damage.

The arcing ground-fault may be as low as 38% of a three-phase bolted short-circuit current. Therefore, the equipment grounding system must be designed to provide an effective ground-fault current path that provides a low-impedance circuit, and that will facilitate the operation of the overcurrent device or ground detector for an ungrounded or impedance-grounded system (250.4(A)(5))(250.4(B)(4)). The effective ground-fault current path mast have sufficient current-carrying capacity to safely carry the ground-fault current that may be imposed on it (solidly grounded system), or to promptly clear the overcurrent devices in the event of a second ground-fault on an opposite phase in an ungrounded or impedance-grounded system.

The various types of equipment grounding conductors are referenced in 250.118. The minimum sizes of equipment grounding conductors are identified in Table 250.122, based on the ampere rating of the circuit overcurrent device. The actual size of the equipment grounding conductor may, at times, have to be increased due to circuit conditions. These conditions include higher levels of ground-fault current and voltage-drop. An effective ground-fault current path must be of sufficiently low impedance, not only to facilitate the operation of the circuit overcurrent device on a solidly grounded system, or to facilitate the operation of the overcurrent devices on an ungrounded or impedance-grounded system in the event of a second ground-fault before the first ground-fault is cleared, but, also to limit the voltage-to-ground, above earth potential, on the equipment grounding system. At times the equipment grounding conductor may be as large as the ungrounded conductor(s). But, in no case is it required to be larger (250.122(A)).

Another consideration is whether or not to install an equipment grounding conductor that is insulated or bare (250.118(1)). Due to the fact that 300.3(B) requires all of the conductors of the same circuit, including the equipment grounding conductor, to be installed in the same raceway or cable, or, in close proximity in the same trench (300.5(I)) for important impedance reduction, it makes sense that an insulated equipment grounding conductor may be a better choice than one (or more) that is bare. This will reduce thermal stress on adjacent conductors during the ground-fault, until this fault is cleared. Even for the direct current conductors of the photovoltaic array circuits, the equipment grounding conductors must be contained within the same raceway, cable or otherwise run with these conductors when these circuit conductors leave the vicinity of the PV array (690.43(C)).

After the ground-fault current has been determined by calculation, the actual size of the equipment grounding conductor will be determined.

To this end, we have provided a detailed example in the next Chapter to identify short-circuit and ground-fault conditions at different locations in a distribution system.

In addition, there are tables that identify short-circuit insulation withstand ratings, as well as the fusing or melting currents for copper conductors from 14 AWG (2.08mm^2) to 500 kcmil (253mm^2), based on time constants ranging from 5 seconds to 1/8 cycle (0.002 seconds).

(Check the UL guide on FHIT cable systems (e.g., cables that are installed on the life/safety and critical branches of an emergency system), where the equipment grounding conductors are installed in raceways and the equipment grounding conductor and raceway is a part of the electrical circuit protection system).

5

A Comprehensive Analysis of a 3-Phase, 4-Wire Distribution System

In this Chapter, we will perform a fault current analysis of a 3-phase distribution system, beginning at the secondary of a transformer, and then downstream to the service equipment, and to a second transformer (separately-derived system), and to the secondary distribution equipment (panelboard), and finally, to a branch-circuit supplying utilization equipment (luminaires).

The short-circuit current at each location will be identified, as well as the ground-fault current, so that the appropriate sizes of the equipment grounding system may be determined.

Transformer 1500kVA- Utility Owned

Voltage – Primary – 26,000 V.

Secondary – 480/277 V. (Wye)

Percent impedance (nameplate) – 3.50%

Note: UL listed transformers, 25kVA and larger have a + or – 10% impedance tolerance, in this case 3.85% or 3.15%.

For two-winding transformers built to ANSI standards, the + or - impedance tolerance is 7.5%.

In our calculation, we will use the worst case of 3.15% -Z (3.50 × .90=3.15).

1. The full – load secondary current is:

$$\frac{1{,}500{,}000\,VA}{480V. \times 1.732} = 1804.27\,or\,1804.3\,amperes$$

2. $\frac{100}{3.15(z)} = 31.746$

3. $1804.3\ amperes \times 31.746 = 57{,}279\ amperes$

57,279 amperes is the short-circuit current that is available at the transformer secondary terminals (bushings). However, if the entire secondary load current is motor load, the available short-circuit current will be increased by a factor of 4 to make up for the rotational energy of this type of load and its effect on the available short-circuit current. In our example, we will assume that 25% of the total load is motor load.

1804.3 amperes × 4 = 7217.2 × 0.25 =1804.3
57,279 amperes + 1804.3 amperes = 59,083.3 amperes

The available short-circuit current at the transformers secondary is now 59,083.3 amperes.

In this example, we have installed 5-750 kcmil-THHN/THWN-2 copper conductors, per phase, with a full size neutral, in nonmetallic conduits (310.10(H)(1)). The major portion of the load is not nonlinear, so the neutral conductors are not considered to be current carrying conductors (310.15(B)(5)(c)).

The nonmetallic conduits are 5″ (metric designator 129), Schedule 80 PVC. Table C.10 of Chapter 9 permits a 4″ (metric designator 103), Schedule 80 PVC conduit for 4-750 kcmil THHN/THWN-2 conductors, but for wire pulling purposes, the conduit size has been increased to 5″.

The length of the secondary conductors, from the transformer secondary to the line side of the service disconnect, which is a 3-pole, 2000 ampere, molded-case circuit breaker, is 100 feet.

1. The calculated short-circuit current at the line terminals of the 3-phase, 2000 ampere, molded-case circuit breaker is as follows:

Voltage-Drop

$$\frac{1.732\,x\,12.9 \times 100\,feet \times 2000\,amperes}{3{,}750{,}000\,cm}$$

$$\frac{4{,}468{,}560}{3{,}750{,}000} = 1.19\,volts$$

$$\begin{array}{r} 480\,Volts \\ -\ 1.19\,volts \\ \hline 478.81, or\ 479\ volts \end{array}$$

479 Volts /1.732 = 276.55, or 277 Volts

Voltage at the service equipment – 479/277 V.
Note: 12.9 = DC resistance for one foot of copper × the conductor circular mil area (NEC Table 8-Chapter 9).

Example

6 AWG copper (uncoated) – 0.491 ohms-1000 feet – Table 8-Chapter 9

$$\frac{0.491}{1000} = 0.000491$$

$$\begin{array}{r} 0.000491 \\ \times 26{,}240\,cm \\ \hline 12.88\,ohms,\ or\,12.9\,ohms \end{array}$$

Note: 21.20-DC resistance for one foot of aluminum x the conductor circular mil area.

6 AWG aluminum = 0.808 ohms-1000 feet – Table 8-Chapter 9

$$\frac{0.808}{1000} = 0.000808$$

$$\begin{array}{r} 0.000808 \\ \times\ 26{,}240\,cm \\ \hline 21.20\,ohms \end{array}$$

Information from ANSI/UL 1581-2011-National Bureau of Standards Handbook 100–1966 and 109-1972

Also, see 210.19(A) Informational Note No.4 and 215.2(A)(1)(b) Informational Note No. 2 for voltage-drop recommended limits.

Short-Circuit Current at Service Disconnect

$$\frac{1.732 \times 100\ feet \times 57{,}279}{29{,}735(750kcmil)(380mm^2) \times 5(per\ phase) \times 479V.} = 0.1393$$

Note: See the tables of "C" values at the end of this Chapter

$$\frac{1}{1 + 0.1393} = 0.8777$$

$$\begin{array}{r} 57{,}279 \\ \times 0.8777 \\ \hline 50{,}273.77,\ or\,50{,}274\,amperes \end{array}$$

$$\begin{array}{rl} 50{,}274.00 & \\ +\ 1804.30 & (motor\ contribution) \\ \hline 52{,}078.30 & amperes \end{array}$$

This short-circuit current establishes the interrupting rating of the 2000 ampere circuit breaker in accordance with 110.9.

Due to the fact that these supply conductors are service conductors, and they are installed in nonmetallic conduits, there will not be a line to equipment ground-fault. However, there may be a bolted line to neutral fault, which may be as high as 100% of the three-phase bolted fault near the transformer (59,083.3 amperes), or 50% of this fault current further downstream.

Section 110.24(A)(B) requires that service equipment, at other than dwelling units, be legibly marked in the field with the maximum available fault current, in this case, 52,078.30 amperes. This field marking shall be of sufficient durability to withstand the environment involved (110.21(A)(1)).

The service equipment must have a short-circuit current rating that is equal to, or greater than the available fault current.

Section 230.90 states that each ungrounded service conductor shall have overload protection. In our example, the service overcurrent device is a 2000 ampere, thermal-magnetic, molded-case circuit breaker, which does not provide overcurrent protection for the service conductors. The ampacity of these conductors is 2375 amperes, per phase, without any correction or adjustment factors applied. This satisfies 230.90(A), as the overcurrent protection is in series with each ungrounded service conductor, and this overcurrent protection does not exceed the allowable ampacity of the service conductors.

The supply transformer in our example is supplied by the serving utility, and therefore, not subject to the transformer overcurrent protection requirements of 450.3(A). And, because this transformer is connected delta-to wye, there will be a 30 degree phase-shift, where the primary voltage leads the secondary voltage by 30 degrees. A fault on the secondary side of the transformer, either line-to-line or line-to-neutral, unless it is a bolted fault within, or very close to the transformer, may not cause the opening of the transformer primary overcurrent protection. And, the service conductors are installed in parallel, so a phase-to-neutral fault in one of these raceways will certainly not produce enough current to open the primary overcurrent protection. Unless the secondary fault occurs at the transformer secondary windings, phase-to-phase or phase-to-neutral, the opening of the primary overcurrent protection is unlikely to occur. The use of cable limiters, in series with each ungrounded secondary conductor is an important consideration.

Another consideration is whether selective coordination is a requirement for this installation. As we have referenced in a previous Chapter, selective coordination is mandatory in several installations. These include elevators (620.62), Critical Operations Data Systems (645.27), Power Source(s) for Electric Motor-Driven Fire Pumps (695.3), Emergency Systems (700.32), Legally Required Standby Systems (701.27), and Critical Operations Power Systems (708.54). Even though there may not be a requirement to have selective coordination, it is certainly a good idea as a safety consideration and as a means of eliminating unnecessary outages, which may have an adverse effect on normal business activities.

In our example, we will include selective coordination, as if it is a requirement.

The 2000 ampere, thermal-magnetic circuit breaker, which is our service disconnecting means has a maximum adjustable instantaneous trip setting of 8 times its rating. An easy method of determining the approximate point where the circuit breaker reaches its instantaneous-trip region is to multiply the circuit breaker rating by its trip setting. In this case, 8 x 2000 amperes, or 16,000 amperes. In addition, there will be an instantaneous-trip pickup tolerance, which must be determined from the manufacturer.

In our example we will use + or – 20%. So, the 2000 ampere circuit breaker may enter its instantaneous trip region at 19,200 amperes (16,000 amperes × 1.20), or 12,800 amperes (16,000 amperes × 0.80). We will use 12,800 amperes for selective coordination purposes.

This service is a 3-phase, 4-wire solidly grounded, wye electrical system with a nominal voltage of 480/277 volts and a service disconnect rated at 2000 amperes. Therefore, Section 230.95 requires ground-fault protection of equipment. As we have stated throughout this book, the most common type of fault in a distribution system is a ground-fault, and the vast majority of these faults are arcing type faults. It is extremely important that the downstream overcurrent devices are selectively coordinated with the ground fault relay. Section 230.95 permits the maximum setting of the ground-fault relay to be 1200 amperes. The 2000 ampere circuit breaker, in consideration of the instantaneous trip setting at 8 times its rating, and, in consideration of the pickup tolerance at –20 % of the circuit breaker rating (12,800 amperes), the time delay of the ground-fault relay must be set to allow low level downstream ground-faults to be cleared by their respective protection devices without opening the 2000 ampere main circuit breaker.

The available short-circuit current at the line terminals of the service disconnect has been calculated at 52,078.30 amperes.

A bolted line-to-ground fault may be 100% of the calculated short-circuit fault near the 2000 ampere circuit breaker, and 50% of this fault current further downstream, or 26,039.15 amperes. And significantly less on the feeders and branch circuits supplied from the service equipment. The 2000 ampere circuit breaker is equipped with a shunt-trip and ground-fault sensor, which, in accordance with 230.95(A), shall have a maximum setting of 1200 amperes and a maximum time delay of one second for ground-fault currents equal to or greater than 3000 amperes.

Because this system is solidly grounded, we will use a metal in-ground support structure, which is in direct contact, vertically, with the earth for 10 feet (3.0m) or more, as the grounding electrode. This support structure is bonded to the metal frame of the building, which is also connected to ground through the rebar in the concrete footings (250.52(A)(2)(3). A copper grounding electrode conductor is extended from the neutral busbar in the service equipment to the grounding electrode system. From Table 250.66, this conductor is 3/0 AWG copper. This conductor is certainly large enough for the supply system, and the most important part of the installation is its length and the method of making the connection to the grounding electrode (system). In order to limit the voltage-rise above earth potential on the connected systems and equipment, the grounding electrode conductor must not be any longer than necessary. Bends and loops in this conductor should be avoided (250.4(A)(1), Informational Note No.1) The currents produced by lightning have an alternating current component that has a frequency range of 3kHz to 10mHz. Frequency is a function of inductive reactance, as $XL = 2\pi FL$. In order to limit the inductive reactance and the overall impedance of the earth connection, the overall length of this conductor and its method of installation is the primary concern. In addition, there will be a grounding electrode (system) at the transformer secondary, connected at the neutral bushing.

The other consideration is the size of the Main Bonding Jumper. This conductor will typically be significantly smaller than the ungrounded service conductors. All of the ground-fault current will flow through this conductor until the appropriate overcurrent device clears this current, or until the ground-fault relay senses the ground-fault and signals the disconnecting means to open.

250.24(B) states that an unspliced Main Bonding Jumper shall be used to connect the equipment grounding conductors(s) and the service-disconnect enclosure to the grounded conductor in accordance with 250.28. The minimum size of the Main Bonding Jumper is in accordance with Table 250.102(C)(1). In our example, the ungrounded service conductors are 3750 kcmil, per phase, copper, and, in accordance with Note 1 of Table 250.102(C)(1), the copper Main Bonding Jumper minimum size will be 12½% of the area of the ungrounded phase conductors, or 12½% of 3750 kcmil.

$$\begin{array}{c} 3{,}750{,}000\,cm \\ \times 0.125 \\ \hline 468{,}750cm\ or\ 500{,}000cm \end{array}$$

The Main Bonding Jumper size is 500kcmil, copper.

In our example, we determined that the available short-circuit current at the line terminals of the service disconnect (2000 ampere, 3-pole, molded-case circuit breaker) is 52,078.30 amperes. The minimum phase-to-ground arcing fault may be 19,789.75, or 19,790 amperes (38% of 52,078.30 amperes). But, even if we base the ground-fault current at 100% of the three-phase bolted short-circuit current, or 52,078.30 amperes, the 500 kcmil copper Main Bonding Jumper will be more than sufficient.

The clearing time of the 2000 ampere molded-case circuit breaker is 0.025 seconds and the fusing current of the 500 kcmil copper Main Bonding Jumper for 0.025 seconds is 436,752 amperes.

A 112.5kVA, 3-phase, 480/208/120 volt transformer is supplied from the service equipment. The transformer marked impedance is 1.24%. The feeder conductors are THHN copper and the conductors are installed in a rigid metal conduit. The length of the feeder is 75 feet and the rigid metal conduit (steel) is connected to the transformer enclosure with a 4 foot length of flexible metal conduit. Because the feeder conductors will be provided with overcurrent protection in excess of 20 amperes, and the trade size will exceed 1¼″ (metric designator 35), the flexible metal conduit is not a type of equipment grounding conductor (250.118(5)(b)(c)) (ANSI/UL1).

250.134(B) normally requires equipment that is fastened in place or connected by permanent wiring methods to be grounded by any of the equipment grounding conductors that are permitted by 250.118. This includes the connection to an equipment grounding conductor that is within the same raceway, cable, or otherwise run with the circuit conductors for important impedance reduction. As the space between conductors increases, magnetic flux density decreases, and the circuit impedance increases. This increase in impedance will reduce the ground-fault current through the equipment grounding conductor, causing the circuit overcurrent device to delay its operation, or not clear at all. An effective ground-fault current path is required to be a low-impedance circuit that facilitates the operation of the circuit overcurrent device and limits the voltage-rise on the metal frames of equipment as the ground-fault is cleared. On an ungrounded or impedance grounded system, the effective ground-fault current path is meant to assure the prompt clearing of the overcurrent devices in the event of a second ground-fault from an opposite phase of the wiring system (250.4(A)(5)(B)(4)).

For a direct current system, the equipment grounding conductor may be run separately from the circuit conductors. In this case, the resistance of this conductor is the important consideration, as the circuit impedance is not a factor (250.134(B), Exception No.2. However, 690.43(C) requires equipment grounding conductors for photovoltaic array circuits and support structures to be contained within the same raceway, cable, or otherwise run with the photovoltaic array circuit conductors. This is an example where a reference in Chapter 6 modifies a reference in Chapter 2 (90.3).

A properly sized external bonding conductor is permitted to supplement the 4 foot length of flexible metal conduit at the transformer (250.102(E)(2)).

This bonding jumper must be routed with the flexible metal conduit and its size is in accordance with 250.122.

However, in our example, the transformer feeder circuit includes an insulated (THHN) copper equipment grounding conductor.

The 112.5kVA transformer has a full-load primary current of 135.60 amperes.

$$\frac{112{,}500\,VA}{479V.\times 1.732} = 135.60\,amperes$$

Table 450.3(B) permits the transformer primary overcurrent protection to be 250% of the full-load primary current.

$$\begin{array}{l} 135.60\,amperes \\ \underline{\times 2.50} \\ 339\,amperes \end{array}$$

However, in our example, the primary overcurrent device will be a 175 ampere circuit breaker (135.60 amperes × 1.25 = 169.50, or 175 amperes).

Based on the use of a 175 ampere, 3-pole molded-case circuit breaker, the feeder conductors appear to be 2/0 THHN copper, which has an ampacity of 175 amperes at 75°C., and this is in accordance with 110.14(C)(1)(b)(2).

The insulated equipment grounding conductor has a minimum size of 6 AWG copper, in accordance with Table 250.122.

But, the size of this conductor will be determined in accordance with the ground-fault current in the feeder circuit. The available short-circuit current at the line terminals of the service disconnect is 52,078.30 amperes. The 175 ampere molded-case circuit breaker for the transformer primary circuit has an adjustable trip which is set at 8 times its rating, and this circuit breaker will enter its instantaneous-trip region at 1400 amperes (175 amperes × 8). The positive and negative tolerance is + or – 20%. So, this circuit breaker may enter its instantaneous-trip region at 1120 amperes (1400 amperes × 0.80). Comparing 1120 amperes with 12,800 amperes for the upstream 2000 ampere circuit breaker, shows that these two overcurrent device are selectively coordinated in their short-circuit regions, and, in most cases they will be selectively coordinated in their overload regions, as well. However, that must be determined by comparing the time-current curves of the circuit breakers, as the unlatching time may vary from several seconds to many minutes (or longer), based on the level of overload current. And there will also be an overload tolerance band.

The available short-circuit current at the line terminals of the 175 ampere circuit breaker is 52,078.30 amperes, as this device is within the service equipment, negating the impedance of the service equipment. So, the interrupting rating of this device must be equal to or greater than this current to satisfy 110.9.

A standard molded-case circuit breaker (600 volts), with no short-time delay feature, has a fault clearing time of 0.025 seconds (1.5 cycles -60Hz-IEEE 1584-Table 1).

Next, we must calculate the short-circuit insulation withstand rating of the 2/0 AWG copper THHN conductors, and the 6 AWG copper THHN equipment grounding conductor.

2/0 AWG copper= 133,100 circular mils
I^2T = one ampere for every 42.25 circular mils for 5 seconds.

$$\frac{133{,}100}{42.25} = 3150\,amperes\ for\ 5\ seconds$$

3150 amperes × 3150 amperes × 5 seconds = 49,612,500 ampere-squared seconds

$$\frac{49{,}612{,}500}{0.025\,seconds} = 1{,}984{,}500{,}000$$

$$\sqrt{1{,}984{,}500{,}000} = 44{,}547.72, or\ 44{,}548\ amperes$$

The insulation withstand rating of the 2/0 AWG copper conductors for 0.025 seconds is 44,548 amperes. And the available short-circuit current is 52,078.30 amperes. This is not acceptable, and a violation of 110.10

Let's try 3/0 AWG (85mm^2) copper THHN conductors.

3/0 AWG copper = 167,800 circular mils

$$\frac{167{,}800}{42.25} = 3971.59, or\ 3972\ amperes\ for\ 5\ seconds$$

3972 amperes × 3972 amperes × 5 seconds = 78,883,920 ampere-squared seconds

$$\frac{78{,}883{,}920}{0.025\,seconds} = 3{,}155{,}356{,}800$$

$$\sqrt{3{,}155{,}356{,}800} = 56{,}172.56, or\ 56{,}173\ amperes$$

The primary feeder conductors are 3/0 AWG THHN copper conductors

The minimum size copper equipment grounding conductor, based on the 175 ampere circuit breaker is determined to be 6 AWG, from Table 250.122. The ground-fault current may vary from 38% of 52,078.30 amperes (minimum) for a line-to-ground arcing fault to 50% of 52,078.30 amperes for a line-to-ground bolted fault. The most common type of ground-fault is the arcing type fault, but, in our example, we will use the 50% value.

$$\begin{array}{r} 52{,}078.30\,amperes \\ \times\,.5 \\ \hline 26{,}039.15\,amperes \end{array}$$

Now, the question becomes whether to use the insulation withstand rating of the 6 AWG copper conductor, or the fusing or melting current of this conductor.

Let's see the difference

Insulation Withstand Rating

6 AWG copper = 26,240 circular mils

$$\frac{26{,}240}{42.25} = 621\,amperes\ for\ 5\,seconds$$

621 amperes × 621 amperes × 5 seconds = 1,928,205 ampere-squared seconds

$$\frac{1{,}928{,}205}{0.025\,seconds} = 77{,}128{,}200$$

$$\sqrt{77{,}128{,}200} = 8782\,amperes\ for\ 0.025\,seconds$$

The insulation withstand rating of the 6 AWG copper conductor for 0.025 seconds is 8782 amperes

Fusing Current

6 AWG, copper = 26,240 circular mils
I^2T = one ampere for every 16.19 circular mils for 5 seconds

$$\frac{26{,}240}{16.19} = 1621\,amperes\ for\ 5\,seconds$$

1621 amperes × 1621 amperes × 5 seconds = 13,138,205 ampere-squared seconds

$$\frac{13{,}138{,}205}{0.025\,seconds} = 525{,}528{,}200$$

$$\sqrt{525{,}528{,}200} = 22{,}924\,amperes$$

The fusing current of the 6 AWG copper conductor is 22,924 amperes for 0.025 seconds

The ground-fault current of 26,039.15 amperes exceeds the insulation withstand rating, and even the fusing current of the 6AWG copper equipment grounding conductor. This conductor does not provide an effective ground-fault current path for this feeder (250.4(A)(5)).

Let's try 4 AWG, copper

Insulation Withstand Rating

4 AWG, copper = 41,740 circular mils

$$\frac{41,740}{42.25} = 988\,amperes\ for\ 5\,seconds$$

988 amperes × 988 amperes × 5 seconds = 4,880,720 ampere-squared seconds

$$\frac{4,880,720}{0.025\,seconds} = 195,228,800$$

$$\sqrt{195,228,800} = 13,972\,amperes\ for\ 0.025\,seconds$$

Fusing Current

4 AWG, copper = 41,740 circular mils

$$\frac{41,740}{16,19} = 2578\,amperes\ for\ 5\,seconds$$

2578 amperes × 2578 amperes × 5 seconds = 33,230,420 ampere-squared seconds

$$\frac{33,230,420}{0.025} = 1,329,216,800$$

$$\sqrt{1,329,216,800} = 36,458\,amperes$$

The fusing current of the 4 AWG, copper conductor is 36,458 amperes for 0.025 seconds

The ground-fault current of 26,039.15 amperes exceeds the insulation withstand rating of the 4 AWG, copper conductor for 0.025 seconds (13,972 amperes). But, the fusing or melting current of this conductor (36,458 amperes) is well above the ground-fault current, and this conductor provides an effective ground-fault current path (250.4(A)(5)).

The feeder conductors from the 175 ampere circuit breaker in the service equipment are 3/0 THHN AWG-copper, and the equipment grounding conductor is 4 AWG THHN, copper.

Table 5 – Chapter 9

The rigid metal conduit (steel) for these conductors will be:
3/0 THHN – 0.2679 square inches x 3 conductors = 0.8037 square inches.
4 AWG THHN – 0.0824 square inches

$$\begin{array}{ll} 0.8037 & \textit{square inches} \\ \underline{0.0824} & \textit{square inches} \\ 0.8861 & \textit{square inches} \end{array}$$

Table 4 – Chapter 9

Rigid metal conduit – 2″-1.363 sq. in. (Article 344)
Flexible metal conduit – 2″-1.307 sq. in. (Article 348)

Voltage-Drop (Transformer Primary)

Assuming 479/277 volts at the service equipment, the voltage-drop in the feeder to the transformer will be:

$$\frac{1.732 \times 12.9 \times 75\,\textit{feet} \times 135.6\,\textit{amperes}}{167{,}800\ (3/0)\,\textit{circular mils}} = \frac{227{,}226}{167{,}800} = 1.354\,\textit{volts}$$

$$\begin{array}{l} 479\,\textit{volts} \\ \underline{-1.354} \\ 477.646\,,\textit{or}\ 478\,\textit{volts} \end{array}$$

Voltage at transformer primary-478/276 Volts (478V. /1.732 =275.98, or 276V.)

$$\begin{array}{l} 478\,\textit{volts} \\ \underline{\times\ .433\ (\textit{Secondary / Primary Voltage Ratio})} \\ 206.974,\ \textit{or}\ 207\,\textit{volts} \end{array}$$

Voltage at transformer secondary-207 Volts-Phase-to-Phase
207V./1.732 = 119.515, or 119.52 Phase-to-Neutral

$$\frac{112{,}500\,VA}{207V. \times 1.732} = 313.78, \textit{or}\ 314\,\textit{amperes} - \textit{Full} - \textit{Load Secondary Current}$$

In our example, the secondary overcurrent device will be a 400 ampere, 3-pole-molded-case circuit breaker, (Table 450.3(B), Note 1), and the secondary conductors will be 500 kcmil THHN, copper (380 amperes @75°C.). The overcurrent protection for these conductors, on the line –side of the 400 ampere circuit breaker is the primary 175 ampere, 3-pole circuit breaker, which may or may not provide adequate protection, due to the delta-to-wye connected transformer.

450.21(A) states that dry-type transformers that are installed indoors and rated at 112.5 kVA, or less shall have a separation of at least 12 inches (300mm) from combustible material, or be separated from the combustible material by a fire-resistant, heat-insulated barrier.

However, an exception waives this requirement for this equipment, where the voltage is 1000 volts, or less, and the transformer is completely enclosed, except for ventilating openings.

The full-load secondary current is 314 amperes (112,500VA/207V. × 1.732), and the secondary conductors will not be protected by the 175 ampere primary circuit breaker, due to the 30 degree phase-shift of the delta connected primary and wye connected secondary (240.4(F)). If the transformer secondary overcurrent protection is rated or set at 125% of the transformer full-load secondary current, this will be 314 amperes × 1.25, or 392.5 amperes.

450.3(B) applies to the overcurrent protection of transformers, and Informational Note No. 1 applies to the overcurrent protection of conductors, and this includes 240.4, 240.21, and 240.100. Therefore, if a 400 ampere nonadjustable circuit breaker or a set of 400 ampere fuses are used as the transformer secondary overcurrent protection, the transformer secondary conductors must be protected in accordance with 240.4(A) through (G), or 240.100(A)(B)(C) for voltages over 1000 nominal.

In our example, the transformer secondary conductors will be 500 kcmil THHN copper, which have a 75°C. ampere rating of 380 amperes, protected at 400 amperes in accordance with 240.4(B). The length of the secondary conductors to the 400 ampere circuit breaker is 25 feet.

The location of the secondary overcurrent device, which in this case, will be a 400 ampere, molded-case, 3-pole, instantaneous-trip circuit breaker, in accordance with 240.21(C)(6).

If the provisions of 240.21(C)(6) are followed, the secondary conductor may be sized as follows:

1. The transformer secondary conductors, from the transformer secondary terminals to the 400 ampere secondary circuit breaker, will have a length of no more than 25 feet (7.5m).

These conductors must have an ampacity of not less than the value of the primary-to-secondary voltage ratio multiplied by one-third of the rating of the transformer primary overcurrent device (175 amperes).

$$\frac{Primary - 478\,volts}{Secondary - 207\,volts} = 2.31$$

$$\frac{175\,amperes}{3} = 58.33\left(\frac{1}{3}of\,175A.\right)$$

$$\begin{array}{r} 2.31 \\ \times 58.33 \\ \hline 134.74\,or\,135\ amperes \end{array}$$

At 135 amperes, with no permission to use the next standard size of over current device, these conductors will be 1/0 copper at 75°C. (150 amperes) to comply with 240.21(C)(6).
The transformer secondary conductors must have an ampacity of no less than 135 amperes.

2. The secondary conductors terminate in a single circuit breaker or set of fuses that limit the load current to no more than the conductor ampacity that is permitted by 310.15. Table 310.15(B)(16) identifies the ampacity of a 500 kcmil THNN copper conductor at 430 amperes (90°C.), and the circuit breaker rating is 400 amperes.
3. The secondary conductors are protected from physical damage by being enclosed in an approved raceway or by other approved means.

In our example, the secondary conductors have an ampacity of 380 amperes, which is the ampacity of the 500 kcmil THHN copper conductors at 75°C. This will be in accordance with 110.14(C)(1)(b)(2), as the terminals of the 400 ampere circuit breaker will be listed and identified for 75°C., unless the terminals are listed and identified for a higher temperature.

The connection from the transformer to a steel junction box will be through a flexible metal conduit. From the junction box to the 400 ampere circuit breaker enclosure, the conductors will bc in a rigid metal conduit (steel).

Once again, the flexible metal conduit is not an equipment grounding conductor in accordance with 250.118(5)(b)(C). An insulated (THHN) copper equipment grounding conductor is installed from the transformer to the 400 ampere circuit breaker enclosure. Its size is in accordance with Table 250.122, based on the 400 ampere circuit breaker rating, or a minimum size of 3 AWG copper.

The available short-circuit current at the primary terminals of the 112.5 kVA transformer will be as follows:

$$\frac{1.732 \times 75\,feet \times 52{,}078.30\,amperes}{12{,}844\ (3/0AWG) \times 478V.} = \frac{6{,}764{,}971}{6{,}139{,}432} = 1.1019$$

$$\frac{1}{1+1.1019} = 0.4758$$

$$\begin{array}{l} 52{,}078.30 \\ \underline{\times 0.4758} \\ 24{,}778.85, or\,24{,}779\,amperes \end{array}$$

The available short-circuit current at the transformer primary terminals is 24,779 amperes.

The available short circuit current at the transformer secondary will be:

$$\frac{24{,}779\,amperes \times 478\,V. \times 1.732 \times 1.24\,Z \times 0.9}{100{,}000 \times 112.5}$$

$$\frac{22{,}894{,}109}{11{,}250{,}000} = 2.0350$$

$$\frac{1}{1+2.0350} = 0.32948, or\,0.3295$$

$$\frac{478 \times 0.3295 \times 24{,}779}{207\,volts} = \frac{3{,}902{,}717}{207\,volts} = 18{,}854\,amperes$$

The available short-circuit current at the transformer secondary terminals is 18,854 amperes.

And, at the line terminals of the 400 ampere circuit breaker, the available short-circuit current will be:

$$\frac{1.732 \times 25\,feet \times 18{,}854\,amperes}{22{,}185 \times 207\,V.}$$

$$\frac{816{,}378}{4{,}592{,}295} = 0.1777$$

$$\frac{1}{1+0.1777} = 0.8491$$

$$\begin{array}{r} 18{,}854\,amperes \\ \times 0.8491 \\ \hline 16{,}009\,amperes \end{array}$$

The interrupting rating of the 400 ampere circuit breaker must be no less than 16,009 amperes (110.9).

The insulation withstand rating of the 500 kcmil copper phase conductors from the transformer to the 400 ampere secondary circuit breaker will be based on the clearing time of the primary 175 ampere circuit breaker which has been determined to be 0.025 seconds. Remember that this transformer is connected delta-to-wye. There will be a 30 degree phase-shift from primary to secondary, as the primary voltage leads the secondary voltage by 30 degrees. This is the reason that 240.4(F) states that the transformer secondary conductors, other than single-phase, 2-wire, and multi-phase, other than delta-to-delta, 3-wire, shall not be considered to be protected by the primary overcurrent protective device. In effect, the transformer secondary conductors require overcurrent protection. And this protection is provided by the secondary 400 ampere circuit breaker. However, the secondary conductors, from the transformer secondary terminals to the 400 ampere circuit breaker are not provided with overcurrent protection by this circuit breaker. The only overcurrent protection for these conductors is the 175 ampere primary overcurrent device, which is installed in the service equipment.

The available short-circuit current at the transformer secondary has been determined to the 18,854 amperes. And, 25 feet of conductor length to the secondary 400 ampere circuit breaker, the available short-circuit current is 16,009 amperes.

If we use 18,854 amperes as the basis for the insulation withstand rating, and 0.025 seconds for the clearing time of the primary 175 ampere circuit breaker, we can calculate the withstand rating of the 500 kcmil copper conductors.

$$\frac{500{,}000\,cm}{42.25} = 11{,}834\,amperes\ for\ 5\,seconds$$

$$11{,}834 \text{ amperes} \times 11{,}834 \text{ amperes} \times 5 \text{ seconds} = 700{,}217{,}780$$

$$\frac{700{,}217{,}780}{0.025\,seconds} = 28{,}008{,}711{,}200$$

$$\sqrt{28{,}008{,}711{,}200} = 167{,}358\,amperes$$

The insulation withstand rating of the 500 kcmil copper conductors for 0.025 seconds is 167,358 amperes, which is well above the available short-circuit current of 18,854 amperes at the transformer secondary terminals. And, even if the secondary fault was a phase-to-neutral bolted fault at, or near, the transformer secondary terminals, and the fault current was 125% of the available fault current (23,568 amperes), the insulation will not be damaged.

As a point of reference, the secondary neutral is full size, or 500 kcmil, copper.

450.10 applies to the bonding and grounding conductor connections inside a dry-type transformer enclosure. There will be a terminal bar bonded to the enclosure, and this bar is not to block any ventilating openings. Continuity between the bar and the enclosure must not be affected by paint or enamel on the enclosure (250.12).

If the dry-type transformer is provided with wire-type connections and not terminal connections, any of the connecting means referenced in 250.8 are acceptable (450.10, Exception).

This dry-type transformer is a separately-derived system, and the bonding connection between the secondary neutral point and the terminal bar or the wire-type connections, is a System Bonding Jumper.

It is sized in accordance with Table 250.102(C)(1), or a 1/0 AWG copper conductor. A 3/0 AWG aluminum or copper-clad bonding jumper is also acceptable. The fusing or melting current of this bonding conductor should be calculated, as well. If a 1/0 AWG copper bonding conductor is used, the fusing or melting current for 0.025 seconds (the clearing time of the circuit breaker) will be:

1/0 AWG -105,600 circular mils

$$\frac{105{,}600\,cm}{16.19} = 6523\,amperes\ for\ 5\,seconds$$

6523 × 6523 × 5 seconds = 212,747,645 ampere-squared seconds

$$\frac{212{,}747{,}645}{0.025\,seconds} = 8{,}509{,}905{,}800$$

$$\sqrt{8{,}509{,}905{,}800} = 92{,}249\,amperes$$

At 92,249 amperes for 0.025 seconds, the 1/0 AWG copper conductor will begin to fuse or melt. The short-circuit current that is available at the transformer secondary is 18,854 amperes. So, the 1/0 AWG copper conductor for the System Bonding Jumper is acceptable. Remember, that a line-to-neutral bolted fault may be as high as 1.25 times this short-circuit current, or 18,854 amperes × 1.25 = 23,568 amperes, if this fault occurs at, or near, the transformer secondary. But, the fusing current of the 1/0 AWG copper bonding jumper is certainly well above this current level.

The secondary of the 112.5kVA transformer is an example of a separately-derived system. The grounding and bonding provisions for this system are derived from 250.30 and 250.102. Based on the size of the 500 kcmil copper ungrounded conductors, the copper grounding electrode conductor will be a minimum size of 1/0 from 250.66. In our example, this conductor will extend from Neutral Point of the wye connected secondary to the grounded metal building frame (250.52(A)(2), which is connected to the earth through a concrete-incased electrode 250.52(A)(3). This conductor will be run as short and straight as practicable (250.4(A)(1), Informational Note No.1).The metal water piping system that is in the area served by the separately-derived system will be bonded to the metal building frame in accordance with 250.104(D)(1), Exception No.2. The size of this bonding jumper will be 1/0 copper from Table 250.102(C) (1).

The equipment grounding conductor run with the secondary conductors to the secondary 400 ampere circuit breaker is an insulated conductor, and its minimum size is determined from Table 250.122, or 3 AWG copper. Once again, if a line-to-ground bolted fault occurs at the transformer, or near the transformer secondary, the available fault current may be 125% of the bolted three-phase fault, or possibly as high as 23,568 amperes. Further downstream, the line-to-ground bolted fault may be 50% of the 3-phase bolted fault, or 18,854 amperes × .50 = 9427 amperes. The ground-fault is, most often, an arcing fault, as opposed to a bolted fault. And, there is a voltage-drop associated with this arcing fault. This will reduce the ground-fault current to, possibly, 38% of the 3-phase bottled fault, or 7165 amperes.

In our example, we will determine the insulation withstand rating of the 3 AWG copper equipment grounding conductor based on a fault current of 23,568 amperes.

3 AWG – 52,620 circular mils

$$\frac{52{,}620\,cm}{42.25} = 1245\,amperes\ for\ 5\,seconds$$

1245 amperes × 1245 amperes × 5 seconds = 7,750,125 ampere-squared seconds

$$\frac{7{,}750{,}125}{0.025\,seconds} = 310{,}005{,}000$$

$$\sqrt{310{,}005{,}000} = 17{,}607\,amperes$$

The insulation withstand rating of the 3 AWG copper conductor is 17,607 amperes for 0.025 seconds.

Based on these conditions, the insulated copper equipment grounding conductor is 1 AWG. However, if the validity of the equipment grounding conductor is based on its fusing current, as opposed to its short-circuit insulation withstand rating, the 3 AWG copper conductor from Table 250.122 is acceptable in providing the effective ground-fault current path required by 250.4(A)(5).

3 AWG – 52,620 circular mils

$$\frac{52{,}620cm}{16.19} = 3250\,amperes\ for\ 5\ seconds$$

3250 amperes × 3250 amperes × 5 seconds = 52,812,500 ampere-squared seconds

$$\frac{52{,}812{,}500}{0.025\,seconds} = 2{,}112{,}500{,}000$$

$$\sqrt{2{,}112{,}500{,}000} = 45{,}962\,amperes$$

In our example, we are considering a fault current that may be as high as 23,568 amperes, at least at the source, and somewhat less downstream. The insulation withstand rating of the 3 AWG copper conductor was determined to be 17,607 amperes for 0.025 seconds. A fault current of 23,568 amperes would certainly destroy this conductor insulation, but this current would be significantly less than the fusing or melting current of this conductor (45,962 amperes vs. 23,568 amperes). 250.4(A)(5) states the effective ground-fault current path shall be capable of safely carrying the maximum ground-fault current likely to be imposed on it from any point on the wiring system where a ground-fault may occur to the electrical supply source. In this case to the transformer secondary windings. And, once again, due to the delta-to-wye transformer connection; and the 30 degree phase-shift associated with this type connection, we are relying on the 175 ampere primary circuit breaker within the service equipment to clear the secondary ground-fault. We have already determined that the 175 ampere circuit breaker, with an adjustable instantaneous-trip setting of 8 times the breaker ampere rating will enter its instantaneous-trip region at 1400 amperes (175 amperes × 8). And, with a tolerance band of + or 20%, the instantaneous-trip region may be as low as 1120 amperes. Due to the delta-to-wye transformer connection, a secondary-fault current of one ampere will produce a primary-fault current of .58 amperes.

If the 3-phase bolted-fault at the transformer secondary was 23,568 amperes, the primary fault current would be 13,669 amperes (23,568 amperes × .58), and this is more than enough current to open the primary 175 ampere circuit breaker.

And, if the fault occurs at the location of the 400 ampere secondary circuit breaker, where the available fault-current has been calculated at 16,009 amperes, even a secondary arcing ground-fault at 38% of the 3-phase bolted-fault (16,009 amperes × .38 = 6083 amperes) will produce enough current to release the tripping mechanism of the 175 ampere circuit breaker and open this circuit in 0.025 seconds.

The secondary feeder from the transformer to the 400 ampere circuit breaker enclosure consists of 4-500 kcmil-THHN copper conductors and a 3 AWG THHN-copper equipment grounding conductor. Even though the 6 foot length of flexible metal conduit is not considered to be an equipment grounding conductor in accordance with 250.118(5) b,c, and ANSI/UL-1 does not recognize flexible metal conduit as a ground-fault return path where the circuit is protected in excess of 20 amperes, or in excess of 6 feet in length, we have provided an internal equipment grounding conductor for the entire circuit length. It should be noted that 250.102(E)(2) recognizes an equipment bonding jumper on the outside of the flexible metal conduit where the length does not exceed 6 feet and the bonding jumper is routed with the flexible raceway.

The total area of the contained conductors is as follows:

$$500\,kcmil - THHN \quad 0.7073\,squ.In. \times 4 = 2.8292\,sq.in.$$

$$3AWG - THHN - 0.0973\,sq.in. \times 1 = \underline{0.0973\,sq.in.} \quad \text{(Table 5 – Chapter 9)}$$

$$2.9265\,sq.in.$$

Flexible metal conduit – 40% fill – 3½″ (metric designator 91) – 3.848 sq. in.
Rigid metal conduit - 40% fill – 3″ (metric designator 78) – 3.000 sq. in.
(Table 4-Chapter 9)

The flexible metal conduit from the transformer secondary to the junction box will be 3½″ trade size.

The rigid metal conduit from the junction box to the 400 ampere circuit breaker enclosure may be 3″ trade size, depending on how the conduit is installed. However, the use of a 3½″ trade size may be better because the feeder conductors have a cross-sectional area that is almost equal to the 40% fill of the 3″ conduit.

<u>Voltage at the 400 ampere circuit breaker will be</u>

$$\frac{1.732 \times 12.9 \times 314A.(FLSC) \times 25\,feet}{500{,}000\,cm}$$

$$\frac{175{,}390.98}{500{,}000} = 0.3508$$

$$\begin{array}{r} 207V. \\ -\ 0.3508V. \\ \hline 206.65V. \end{array}$$

$$206.65\text{v.}/1.732 = 119.31\,\text{volts, at 400 ampere CB}$$

From the 400 ampere circuit breaker, the feeder conductors extend for a length of 100 feet to a 400 ampere panelboard. This circuit breaker satisfies 408.36. It protects the panelboard, as well as the supply conductors, so the panelboard does not require a main overcurrent protective device. The 400 ampere circuit breaker is on the secondary side of the transformer in accordance with 408.36(B), and a 3½″ trade size rigid metal conduit is used.

The available fault current at the 400 ampere circuit breaker is 16,009 amperes. And, the available fault current at the downstream panelboard will be as follows:

$$\frac{1.732 \times 100\,feet\,16{,}009\,amperes}{22{,}185(500kcmil)(253mm^2) \times 207V} = \frac{2{,}772{,}759}{4{,}592{,}295} = 0.6038$$

$$\frac{1}{1 + 0.6038} = 0.6235$$

$$\begin{array}{r} 16{,}009\,amperes \\ \times\ 0.6235\quad \\ \hline 9982\,amperes \end{array}$$

The short-circuit current that is available at the line terminals of the panelboard is 9982 amperes.

Note, there is no motor contribution specified here. If the entire load consisted of motors, the full-load secondary current would be multiplied by a factor of 4, and this value is added to the calculated short-circuit current.

$$\begin{array}{r} 9982\,amperes \\ +1256\ amperes \\ \hline 11{,}238\,amperes \end{array}\ (314\,amperes \times 4)$$

In our example, there is no motor contribution.

Example

In order to calculate the neutral current of a 3-phase, 4-wire, Wye connected supply system, the neutral current can be calculated as shown in the following example:

Phase A – 220 amperes

Phase B – 240 amperes
Phase C – 210 amperes

In this example, the major portion of the neutral connected load is not nonlinear. So, the neutral conductor is not considered to be a current carrying conductor (310.15(B)(5)(C)).

$$I_N = \left(L_{1^2} + L_{2^2} + L_{3^2}\right) - \left[\left(L_1 \times L_2\right) + \left(L_2 \times L_3\right) + \left(L_1 \times L_3\right)\right]$$

$$I_N = \left(48{,}400 + 57{,}600 + 44{,}100\right) - \left[\left(52{,}800\right) + \left(50{,}400\right) + \left(46{,}200\right)\right]$$

$$150{,}100 - 149{,}402$$

$$\sqrt{700} = 26.46\,amperes\,or\,27\,amperes$$

The feeder neutral carries 27 amperes, based on this unbalanced load, in this example.

The 3 AWG THHN copper equipment grounding conductor has are insulation withstand rating of 17,607 amperes for 0.025 seconds, which is the clearing time of the 400 ampere circuit breaker. The insulation withstand rating of this conductor is in excess of the available short-circuit current of 16,009 amperes. And, of course, the fusing current of this conductor for 0.025 seconds is 45,962 amperes. This equipment grounding conductor provides an effective ground-fault current path on its own, without even considering the rigid metal conduit that is in parallel with the 3 AWG copper conductor.

As we stated before, there is no main circuit breaker in the panelboard, but the individual feeder and branch-circuit breakers in the panelboard should be selectively coordinated with the 400 ampere circuit breaker.

For example, consider a 100 ampere circuit breaker in this panelboard that has a fixed instantaneous trip, that ranges from 800-1300 amperes. In our example, this circuit breaker will easily coordinate with the upstream 400 ampere circuit breaker.

However, the overload region must be checked to assure the selective coordination, as well. Also, the interrupting rating of the panelboard circuit breakers must be at least 9982 amperes to satisfy 110.9.

The line-to-neutral bolted fault current at the transformer secondary terminals may exceed the three-phase bolted fault current of 18,854 amperes by 25%. So, our calculation of this line-to-neutral fault will be:

$$\begin{array}{l} 18{,}854\,amperes \\ \underline{\times 1.25} \\ 23{,}568\,amperes \end{array}$$

The length of the secondary conductors to the 400 amperes, 3-pole circuit breaker is 25 feet (7.62 meters). The secondary conductors are 500kcmil, including the secondary neutral conductor.

The line-to-neutral bolted fault current at the 400 ampere-circuit breaker will be:

$$\frac{1.732 \times 25\,feet \times 23{,}568\,amperes}{22{,}185(500kcmil) \times 119V.} = \frac{1{,}020{,}494}{2{,}640{,}015} = 0.3865$$

$$\begin{array}{r} 23{,}568\,amperes \\ \times 0.3865 \\ \hline 9{,}109\,amperes \end{array}$$

And, the line-to-neutral bolted fault current at the 400 ampere panelboard will be:

$$\frac{1.732 \times 100\,feet \times 9{,}109\,amperes}{22{,}185(500kcmil) \times 119V.} = \frac{1{,}577{,}679}{2{,}640{,}015} = 0.5976$$

$$\frac{1}{1 + 0.5976} = 0.6259$$

$$\begin{array}{r} 9{,}109\,amperes \\ \times 0.6259 \\ \hline 5{,}701\,amperes \end{array}$$

Voltage at Panelboard

$$\frac{1.732 \times 12.9 \times 314\,amperes \times 100\,feet}{500{,}000cm}$$

$$\frac{701{,}564}{500{,}000} = 1.4V. - Phase - to - Phase$$

$$\begin{array}{r} 206{,}65V. \\ 1.40V. \\ \hline 205.25V., or\,205V. \end{array}$$

205V./1.732 = 118.36 (phase-to-neutral)

A 15 ampere branch circuit breaker in the 400 ampere panelboard supplies a row of fluorescent lighting fixtures (luminaires). This lighting load is continuous, so the maximum load on this 15 ampere circuit breaker is 12 amperes, that is, 80% of 15 amperes, in accordance with 210.20(A).

The branch circuit conductors are 14 AWG THHN, copper, and, the total length of these conductors is 80 feet (24.38 meters), including the conductor length through the connected luminaires. There are 10 - 4 foot (1.219 meters) fluorescent fixtures connected end-to-end. The branch circuit conductors are extended through a ½″ electrical metallic tubing (steel), and the equipment grounding conductor is 14 THHN copper (250.122) (250.119).

And now, the branch circuit supplying the fluorescent lighting load. The total load of the 10 luminaires is 8.33 or 9 amperes.

$$\text{Ballast} - \frac{100 \text{ VA} \left(80\% \, power \ factor\right)}{120 \text{ V.}} = .833 \, amperes$$

$$\begin{array}{c} .833 \text{ amperes} \\ \underline{\times 10 \text{ luminaries}} \\ 8.33, or\, 9\, amperes \end{array}$$

The total lighting load is 9 amperes. The branch circuit conductors are sized at 125% of the lighting load (1.25 × 9 amperes = 11.25 amperes) to satisfy 210.19(A)(1)(a). And, the 14 AWG copper conductor is protected in accordance with 240.4(D)(3) at 15 amperes. There is no temperature correction factor applied, as the ambient temperature does not exceed 30°C. (86°F).

The voltage-drop at the first luminaire will be as follows (40 feet):

$$\frac{2 \times 12.9 \times 40 \, feet \times 9 \, amperes}{4110 cm \ (14 AWG)} = 2.26V.$$

<u>Voltage at First Luminaire</u>

$$\begin{array}{c} 118.36V. \\ \underline{-2.26V.} \\ 116.10V. \end{array}$$

Based on these conditions, the line-to-neutral bolted fault current at the first luminaire will be as follows:

$$\frac{2 \times 40 \, feet \times \ 5701 \, amperes}{389 \left(14 AWG\right) \times 1 \times 116.10V.} = \frac{456{,}080}{45{,}163} = 10.0985$$

$$\frac{1}{1+10.0985} = 0.0902$$

$$\begin{array}{r} 5701\,amperes \\ \times 0.0902 \\ \hline 514\,amperes \end{array}$$

$$\begin{array}{r} 5701\,amperes \\ -514\,amperes \\ \hline 5187\,amperes \end{array}$$

The voltage-drop at the last luminaire will be as follows:

$$\frac{2\times 12.9\times 80\,feet\times 9\,amperes}{4110cm} = 4.52V.$$

Voltage At Last Luminaire

$$\begin{array}{r} 118.36V. \\ -4.52V. \\ \hline 113.84V. \end{array}$$

And, the line-to-neutral bolted fault current at the last luminaire will be:

$$\frac{2\times 80\,feet\times 5187\,amperes}{389(14\,AWG)\times 1\times 113.84\,V.} = \frac{829{,}920}{44{,}284} = 18.74$$

$$\frac{1}{1+18.74} = 0.0506$$

$$\begin{array}{r} 5{,}187\,amperes \\ \times 0.0506 \\ \hline 262.46\,amperes \end{array}$$

This is significant because, according to the listing instructions for the ballasts of the fluorescent lights, these units must be able to withstand 200 amperes of fault current, and, with the calculated fault currents for this branch circuit, there is a problem with the provisions of 110.10, which states that 'listed equipment applied in accordance with their listing shall be considered to meet the requirements of this section'.

These luminaries must have supplementary overcurrent protection (240.10), typically in the form of fuses, that are within the luminaries, and that provide proper fault current protection for the ballasts (UL1077). These devices are not intended for branch-circuit protection.

The ballasts are also required to have integral thermal protection in accordance with 410.130(E)(1).

We selected a 14 AWG THHN copper conductor for this branch circuit, protected with a 15 ampere circuit breaker (Table 310.16(B) 240.4(D)(3)). And, based on these circuit conditions, this appears to be acceptable.

However, the line-to-neutral fault current has been determined to be 5701 amperes at the 400 ampere panelboard. The 15 ampere molded-case circuit breaker has a fault clearing time of 0.025 seconds. If the line-to-neutral fault developed near this panelboard, will the insulation on the 14 AWG copper conductor be damaged? If so, there is a problem with 110.10.

14 AWG copper - 4110 circular mils

$$\frac{4110\,cm}{42.25} = 97\,amperes\;for\;5\,seconds$$

97 amperes × 97 amperes × 5 seconds = 47,045 amperes squared seconds

$$\frac{47{,}045\,cm}{0.025\,seconds} = 1{,}881{,}800$$

$$\sqrt{1{,}881{,}800} = 1372\,amperes\;for\;0.025\,seconds$$

The 14 AWG THHN copper conductor will withstand 1372 amperes for 0.025 seconds without insulation damage. But, the available fault current, at least near the panelboard, will be approximately 5701 amperes. The insulation of the 14 AWG conductor will certainly be destroyed under these conditions.

The choice will be to increase the conductor size to increase the insulation withstand rating. And, this increase will be significant under these conditions.

In order to satisfy these conditions the branch circuit conductors would have to be increased to 6 AWG, copper (8782 amperes for 0.025 seconds). And, this would make no sense. And it would be a violation of UL 489, which is the listing for the molded-case circuit breaker, because the listing instructions limit the wire size on the 15 ampere circuit breaker to no more than a 2 wire size increase. That is, from 14 AWG to 10 AWG for the 15 ampere circuit breaker.

Using a larger wire than 10 AWG would affect the time-current curves of the 15 ampere circuit breaker due to the heat-sink effect of the larger wire.

The other solution is to use a current-limiting circuit breaker, either for the branch-circuit or ahead of the branch-circuit, such as the 400 ampere feeder circuit breaker.

However, even using a current-limiting circuit breaker for the branch-circuit, with a fault-clearing time of 0.008 seconds, would still necessitate the use of a 8 AWG copper conductor for this branch circuit, which is too large for the 15 ampere circuit breaker, based on the listing instructions.

Reducing the fault current through the use of current-limiting overcurrent devices for the 400 ampere feeder and the 15 ampere branch circuit may be a viable solution.

This is an example where the branch-circuit appeared to be properly protected in accordance with Table 310.16(B) and 240.4(D)(3). But, there is a problem with 110.10, because the short-circuit current rating of the conductor insulation is well below the available fault current.

The 14 AWG, THHN copper equipment grounding conductor may be acceptable, based on its fusing current.

14 AWG copper – 4110 circular mils

$$\frac{4110\,cm}{16.19} = 254\,amperes$$

254 amperes × 254 amperes × 5 seconds = 322,580

$$\frac{322{,}580}{0.008\,seconds\left(\frac{1}{2}cycle\,60Hz\right)} = 6350\,amperes$$

Using the fusing current of the 14 AWG copper equipment grounding conductor, and the use of a current-limiting overcurrent device, with a fault-clearing time of 0.008 seconds, this conductor has a fusing current of 6350 amperes, which is above the fault current of 5701 amperes.

So, this analysis includes the pertinent information for the entire distribution system, from the utility transformer to the farthest piece of utilization equipment supplied by a branch-circuit.

The following tables are derived from IEEE 241 and IEEE 242. These values are equal to one over the impedance per foot of the conductor. The values specified in these tables are used in the short circuit calculations of Chapter 5.

And using the 'C' values that are expressed on the following pages, and the examples that we used in our analysis of the service, feeder, and branch circuit, it is possible to calculate the fault currents from the source to the farthest outlet or utilization equipment.

Conductors & Busways "C" Values

Copper												
	Three Single Conductors						Three – Conductor Cable					
AWG or	Conduit (Steel)			Nonmagnetic			Conduit (Steel)			Nonmagnetic		
kcmil	600V	5kV	15kV	600V	5kV	15kV	600V	5kV	15kV	600V	5kV	15kV
14	389	-	-	389	-	-	389	-	-	389	-	-
12	617	-	-	617	-	-	617	-	-	617	-	-
10	981	-	-	982	-	-	982	-	-	982	-	-
8	1557	1551	-	1559	1555	-	1559	1557	-	1560	1558	-
6	2425	2406	2389	2430	2418	2407	2431	2425	2415	2433	2428	2421
4	3806	3751	3696	3826	3789	3753	3830	3812	3779	3838	3823	3798
3	4774	4674	4577	4811	4745	4679	4820	4785	4726	4833	4803	4762
2	5907	5736	5574	6044	5926	5809	5989	5930	5828	6087	6023	5958
1	7293	7029	6759	7493	7307	7109	7454	7365	7189	7579	7507	7364
1/0	8925	8544	7973	9317	9034	8590	9210	9086	8708	9473	9373	9053
2/0	10755	10062	9390	11424	10878	10319	11245	11045	10500	11703	11529	11053
3/0	12844	11804	11022	13923	13048	12360	13656	13333	12613	14410	14119	13462
4/0	15082	13606	12543	16673	15351	14347	16392	15890	14813	17483	17020	16013
250	16483	14925	13644	18594	17121	15866	18311	17851	16466	19779	19352	18001
300	18177	16293	14769	20868	18975	17409	20617	20052	18319	22525	21938	20163
350	19704	17385	15678	22737	20526	18672	22646	21914	19821	24904	24126	21982
400	20566	18235	16366	24297	21786	19731	24253	23372	21042	26916	26044	23518
500	22185	19172	17492	26706	23277	21330	26980	25449	23126	30096	28712	25916
600	22965	20567	17962	28033	25204	22097	28752	27975	24897	32154	31258	27766
750	24137	21387	18889	29735	26453	23408	31051	30024	26933	34605	33315	29735
1,000	25278	22539	19923	31491	28083	24887	33864	32689	29320	37197	35749	31959

Aluminum

AWG or	Three Single Conductors						Three – Conductor Cable					
	Conduit (Steel)			Nonmagnetic			Conduit (Steel)			Nonmagnetic		
kcmil	600V	5kV	15kV	600V	5kV	15kV	600V	5kV	15kV	600V	5kV	15kV
14	237	-	-	237	-	-	237	-	-	237	-	-
12	376	-	-	376	-	-	376	-	-	376	-	-
10	599	-	-	599	-	-	599	-	-	599	-	-
8	951	950	-	952	951	-	952	951	-	952	952	-
6	1481	1476	1472	1482	1479	1476	1482	1480	1478	1482	1481	1479
4	2346	2333	2319	2350	2342	2333	2351	2347	2339	2353	2350	2344
3	2952	2928	2904	2961	2945	2929	2963	2955	2941	2966	2959	2949
2	3713	3670	3626	3730	3702	3673	3734	3719	3693	3740	3725	3709
1	4645	4575	4498	4678	4632	4580	4686	4664	4618	4699	4682	4646
1/0	5777	5670	5493	5838	5766	5646	5852	5820	5717	5876	5852	5771
2/0	7187	6968	6733	7301	7153	6986	7327	7271	7109	7373	7329	7202
3/0	8826	8467	8163	9110	8851	8627	9077	8981	8751	9243	9164	8977
4/0	10741	10167	9700	11174	10749	10387	11185	11022	10642	11409	11277	10969
250	12122	11460	10849	12862	12343	11847	12797	12636	12115	13236	13106	12661
300	13910	13009	12193	14923	14183	13492	14917	14698	13973	15495	15300	14659
350	15484	14280	13288	16813	15858	14955	16795	16490	15541	17635	17352	16501
400	16671	15355	14188	18506	17321	16234	18462	18064	16921	19588	19244	18154
500	18756	16828	15657	21391	19503	18315	21395	20607	19314	23018	22381	20978
600	20093	18428	16484	23451	21718	19635	23633	23196	21349	25708	25244	23295
750	21766	19685	17686	25976	23702	21437	26432	25790	23750	29036	28262	25976
1,000	23478	21235	19006	28779	26109	23482	29865	29049	26608	32938	31920	29135

To Convert US Customs Units of Measurement to Metric Sizes

Multiply the US Measurement by 25.4, or

Divide the Metric Size by 25.4 to convert to US Measurement

To convert circular mils to metric wire sizes, divide the circular mil area by 1973.53, or multiply the metric wire size by 1973.53 to convert to circular mils.

Inch	=	0.254 meters
Inch	=	2.54 centimeters
Inch	=	25.40 millimeters
Meter	=	39.37 inches
Millimeter	=	0.03937 inch
Centimeters	=	inches/2.54
Foot	=	.3048 meter
Yard	=	0.9144 meters
Mile	=	1609 meters
Kilometer	=	0.6213 miles
Square meter	=	square foot ÷ 0.093
Circumference	=	πd
Area of circle	=	πr^2
Celsius to Fahrenheit	=	temperature × 1.8 + 32
Fahrenheit to Celsius	=	temperature – 32 × 0.5555
Square feet to square meter	=	$m^2 = \frac{ft^2}{10.764}$
π	=	3.1416
$\sqrt{2}$	=	1.414
$\sqrt{3}$	=	1.732
Voltage-drop	=	$\frac{2K \times L \times I}{CM}$ - single-phase
Voltage-drop	=	$\frac{1.732K \times L \times I}{CM}$ - three-phase
Circular mils	=	$\frac{2K \times L \times I}{VD}$ - single-phase
Circular mils	=	$\frac{1.732K \times L \times I}{VD}$ - three-phase

K = 12.90 ohms – copper

K = 21.20 ohms – aluminum

L = one way length in feet of conductor

I = amperes of load
CM = circular mil area of conductor (NEC Table 8-Chapter 9)
Neutral current in a 3-phase, 4-wire, Wye system

$$\sqrt{L1^2 + L2^2 + L3^2 - \left[(L1\times L2) - (L2\times L3) - (L1\times L3)\right]}$$

Series Circuits

Total Resistance - $R_T = R_1 + R_2 + R_3 + R_4$
Resistance Where Voltage and Power (Wattage) are Known

$$R = E^2 P$$

Power (Wattage) - $P = I^2 \times R$
Voltage Equals the Sum of all of the Power Supplies

$$V_T = V_1 + V_2 + V_3 + V_4$$

Current Equals- I = E (Source) / R (Total)

Parallel Circuits

The voltage – drop across each resistance is equal to the voltage supplied by power source.

The current in each branch of the parallel circuit is calculated by the formula $- I = \frac{E}{R}$

Parallel Circuit Power (Wattage) in each branch equals $P = I^2 R$ or $P = E \times I$ or $P = \frac{E^2}{R}$

Total Power - The sum of the power in a parallel circuit equals the sum of the power in each branch.

If all of the resistors have the same resistance, the total resistance equals the resistance of one resistor, divided by the total number of resistors in parallel.

Example

What is the total resistance of 5 – 10 ohm resistors in parallel?

$$\frac{10\,ohms}{5\,resistors} = 2\,ohms$$

Where the paralleled resistors have different ohmic ratings, the total resistance may be calculated using the Product Over Sum Method

Example

Two resistors in parallel – 5 ohms – 7 ohms

$$\begin{array}{r} 5\,ohms \\ \times 7\,ohms \\ \hline 35\,ohms \end{array} \qquad \begin{array}{r} 5\,ohms \\ +7\,ohms \\ \hline 12\,ohms \end{array}$$

$$\frac{35\,ohms}{12\,ohms} = 2.92\,ohms$$

Total resistance = 2.92 ohms

If a third resistor is added (fourth, fifth, sixth, etc.), the same formula may be used to calculate the total resistance.

Example

Two resistors are added, one is 10 ohms and the second is 15 ohms, what is the total resistance?

$$\begin{array}{r} 2.92\,ohms \\ \times 10\,ohms \\ \hline 29.2\,ohms \end{array} \qquad \begin{array}{r} 2.92\ ohms \\ +10\,ohms \\ \hline 12.92\ ohms \end{array}$$

$$\frac{29.2\ ohms}{12.92\ ohms} = 2.26\,ohms$$

$$\begin{array}{r} 2.26\,ohms \\ \times 15\,ohms \\ \hline 33.9\,ohms \end{array} \qquad \begin{array}{r} 2.26\ ohms \\ +15\,ohms \\ \hline 17.26\ ohms \end{array}$$

$$\frac{33.9\ ohms}{17.26\ ohms} = 1.96\,ohms$$

Total resistance = 1.96 ohms

The Reciprocal Method may be used to calculate the total resistance where the resistances have different ratings.

Example

Four resistors in parallel with ratings of 10 ohms – 6 ohms – 8 ohms – 15 ohms

$$\frac{1.00}{\frac{1}{10}+\frac{1}{6}+\frac{1}{8}+\frac{1}{15}}$$

$$\frac{1}{10}=0.1-\frac{1}{6}=0.167-\frac{1}{8}=0.125-\frac{1}{15}=0.067$$

$$\begin{array}{r} 0.100 \\ +0.167 \\ +0.125 \\ +0.067 \\ \hline 0.459 \end{array} \qquad \frac{1.00}{0.459}=2.178\,ohms$$

Total resistance equals 2.178 ohms

Overcurrent Protection

The following 50 questions cover a wide range of topics in the application of overcurrent protection referenced in the National Electrical Code.

Carefully review each question and select the answer before checking the answer key. Review the appropriate NEC section and read any Exception(s) and Informational Note(s).

Good luck, and as always, I welcome your comments and suggestions.

Gregory P. Bierals
Electrical Design Institute

1. 10 foot tap conductors may be protected at up to ________ times the tap conductor ampacity.
   a) 5
   b) 8
   c) 10
   d) 12

2. The term 'overcurrent' is defined as a condition produced by an overload, short-circuit, or ______________.
   a) ground-fault
   b) starting current
   c) magnetizing current
   d) interrupting current

3. A two-pole circuit breaker shall not be used to protect a corner-grounded delta system, unless it is marked ________.
   a) 3-phase only
   b) 2-phase
   c) 1-phase-3-phase
   d) corner-grounded

4. Circuit breakers have a minimum interrupting rating of _____amperes, unless otherwise marked.
   a) 10,000
   b) 5,000
   c) 22,000
   d) 14,000

5. A circuit protected at 400 amperes shall have a minimum aluminum equipment grounding conductor size of _________ AWG
   a) 4
   b) 1/0
   c) 2
   d) 1

6. What is the minimum size of copper feeder tap conductors (THHN), no longer than 25 feet, where the upstream overcurrent protection is rated at 600 amperes.
   a) 2/0
   b) 3/0

c) 1/0
d) 250 kcmil

7. An (overcurrent) device is not normally permitted in series with a grounded (neutral) conductor.
   a) True
   b) False

8. Motor circuit conductors that supply a single-phase motor used in a continuous-duty application shall have an ampacity of not less than _______________percent of the full-load current rating and may be protected at up to _______________ percent of the conductor ampacity by a time-delay fuse.
   a) 110 – 200
   b) 125 – 175
   c) 110 – 120
   d) 120 – 150

9. Where circuit breakers are intended to interrupt current in the normal performance of the function for which they are installed in a Class I, Division 2 Hazardous (Classified) Location, they shall be provided with enclosures indentified for Class I, Division 1.
   a) True
   b) False

10. GFCI protection is required for outlets not exceeding _______________ volts that supply boat hoists installed in dwelling unit locations.
    a) 120
    b) 240
    c) 120/240
    d) b or c

11. What is the maximum rated overcurrent device for a 10 AWG copper, THHN conductor used for an arc welder branch circuit?
    a) 30 A
    b) 45 A
    c) 50 A
    d) 60 A

12. Selective coordination is required for the overcurrent devices supplying fire pump circuits.
    a) True
    b) False

13. For transformers 1000 volts and less, and where the rated primary current is less than 9 amperes, the primary overcurrent protection may be rated at ______________percent of the primary full-load current rating where both primary and secondary overcurrent protection are provided.
    a) 167
    b) 300
    c) 250
    d) 125

14. Potential transformers shall be supplied by a circuit that is protected by standard overcurrent devices rated at ____________ amperes or less.
    a) 15
    b) 20
    c) 10
    d) 25

15. Photovoltaic systems operating at __ volts dc or greater between any 2 conductors shall be protected by a listed PV arc-fault circuit interrupter or be provided with equivalent protection.
    a) 125
    b) 100
    c) 60
    d) 80

16. A type ________________ fuse adapter is designed so that once inserted into a fuseholder it cannot be removed.
    a) P
    b) S
    c) H
    d) E

17. Where motor control circuit conductors extend beyond the controller enclosure, 12 AWG copper conductors may be protected at up

to ______________ amperes by the motor branch-circuit overcurrent device.
a) 45
b) 60
c) 90
d) 70

18. Individual single-pole circuit breakers with ______________ handle-ties shall be permitted to protect the ungrounded conductors of multiwire branch circuits that only serve line-to-neutral loads.
a) approved
b) listed
c) identified
d) certified

19. For feeders operating at over 1000 volts, the continuous ampere rating of a fuse may not exceed ________________ times the ampacity of the conductors.
a) 3
b) 6
c) 4
d) 8

20. Overcurrent devices for emergency systems shall be ________________ with all supply side overcurrent devices.
a) properly rated
b) selectively coordinated
c) connected in series
d) none of these

21. Circuit conductors that supply the power conversion equipment for an adjustable-speed drive system shall have an ampacity of not less than ________________ of the rated input current to the power conversion equipment.
a) 175 %
b) 150%
c) 125 %
d) 100%

22. The overcurrent protection for a hermetic motor compressor may not exceed ________________ of the motor-compressor rated-load current or branch-circuit selection current, whichever is greater.
    a) 175 %
    b) 225 %
    c) 150 %
    d) 125 %

23. 15 ampere and 20 ampere, 125 volt receptacle outlets that are within ________________ of the inside walls of a pool shall be GFCI protected.
    a) 6
    b) 8
    c) 10
    d) 20

24. 18 AWG fixture wire can be tapped to a 20 ampere branch-circuit for a length up to ______ feet.
    a) 100
    b) 50
    c) 25
    d) 15

25. Supplementary fuses, such as Type GLR or GMF, used to isolate fluorescent lights in the event of an overcurrent, may be used as a substitute for the branch-circuit overcurrent device.
    a) True
    b) False

26. A single-phase, 240/120 volt, 50 kVA transformer may have a primary overcurrent device rated at ________ amperes.
    a) 260
    b) 300
    c) 350
    d) 250

27. The transformer in question # 26 does not require overcurrent protection for the secondary conductors if the secondary conductor ampacity is ________________ amperes.
    a) 800

b) 300
c) 400
d) 600

28. Each ungrounded service conductor shall have ____________________ protection.
a) ground-fault
b) short-circuit
c) overload
d) overcurrent

29. Where circuit breakers are back-fed, they shall be ________________ by an additional fastener that requires more than a pull to release.
a) shunt-trapped
b) external mounted
c) secured in place
d) none of these

30. Panelboards equipped with snap switches rated 30 amperes or less shall have overcurrent protection not exceeding.
a) 60A
b) 100A
c) 150A
d) 200A

31. Class H cartridge fuses of the renewable type shall be permitted to be used only for replacement in existing installations where there is no evidence of overfusing or tampering.
a) True
b) False

32. Circuit breakers with slash ratings, such as 480Y/277 V, are permitted to be applied in a system where the voltage-to-ground does not exceed ________________.
a) 120 V
b) 240 V
c) 277 V
d) 300 V

33. The overcurrent protective device for X-Ray equipment shall not be less than ____________ percent of the equipment momentary rating, or ________________ percent of the long time rating, whichever is greater.
    a) 25 – 50
    b) 50 – 100
    c) 75 – 100
    d) 100 – 125

34. A modular data center that directly connects to a service shall have a __________ that is not less than the ______________ of the service.
    a) ampere rating-short-time rating.
    b) short-circuit current rating – available fault current
    c) full-load current- full-load current
    d) a and c

35. Supplementary overcurrent protective devices __________________ be required to be marked with an interrupting rating.
    a) shall
    b) shall not
    c) may
    d) may not

36. One method of arc energy reduction where fuses are rated 1200 amperes or higher is to have a clearing time of ____________ seconds or less at the available arcing current.
    a) 0.04
    b) 0.08
    c) 0.07
    d) 0.10

37. The screw shell of a plug-in type fuseholder shall be connected to the __________ side of the circuit.
    a) neutral
    b) line
    c) ground
    d) load

38. The ballast of fluorescent luminaires installed indoors shall have ________________

a) integral thermal protection
b) external thermal protection
c) overload protection
d) a or b

39. The Class 2 circuit supplying a low-voltage suspended ceiling power distribution system shall be protected at not greater than ___________ amperes.
 a) 10
 b) 15
 c) 25
 d) 20

40. Which of the following is not a standard size fuse
 a) 35A
 b) 45A
 c) 75A
 d) 90A

41. For existing recreational vehicle sites with electric supply, at least 20 percent shall be equipped with a 125/250 volt _______ ampere receptacle.
 a) 20
 b) 30
 c) 40
 d) 50

42. Service equipment for a mobile home shall be rated at not less than __________at 120/240 volts.
 a) 100 amperes
 b) 50 amperes
 c) 60 amperes
 d) 150 amperes

43. Where variable loads are supplied by a phase converter, overcurrent protection shall be set at not more than _________________ of the phase converter nameplate single-phase input full-load amperes.
 a) 115 %
 b) 125 %
 c) 150 %
 d) 200 %

44. Optional standby system equipment shall be suitable for the maximum available short-circuit current at its terminals.
    a) True
    b) false

45. Swimming pool pump motors connected to 120/240 volt branch circuits shall be provided with ________________ protection for personnel.
    a) AFCI
    b) GFCI
    c) a and b
    d) a or b

46. The overcurrent protection for nonpower limited fire alarm circuits (NPLFA) shall not exceed ________________ amperes for 18 AWG conductors and __________ amperes for 16 AWG conductors.
    a) 5 and 8
    b) 6 and 10
    c) 7 and 10
    d) 7 and 15

47. The branch circuit supplying fire alarm equipment shall be supplied through ground-fault circuit interrupters or arc-fault circuit interrupters.
    a) True
    b) False

48. Shore power for boats shall be provided by single receptacles rated not less than ________________ amperes.
    a) 60
    b) 20
    c) 50
    d) 30

49. Grounded dc photovoltaic arrays shall be provided with identified direct-current __________ for the purpose of reducing fire hazards.
    a) ground-fault protection
    b) arc-fault protection
    c) GFPE
    d) ground-fault monitors

50. A branch-circuit overcurrent device may be used for the overcurrent protection for _______________
    a) branch-circuits
    b) feeders
    c) service conductors
    d) all of these

Answer key

1. c - 240.21(B)(1)(4)
2. a - Article 100 – Definition – Overcurrent
3. c - 240.85
4. b - 240.83(C)
5. d - Table 250.122
6. b - 240.21(B)(2)(1) – Table 310.15(B)(16)
7. b - 240.22(1)(2)
8. b - 430.22 – Table 430.52
9. a - 501.115(B)(1)
10. d - 210.8(C)
11. d - 630.12(A)(B) – Table 310.15(B)(16)
    240.4(D) – 240.4(G)
    10AWG – Copper @60°C. -30A.
12. b - 695.5(B)
13. c - Table 450.3(B)
14. a - 450.3(C)-408.52
15. d - 690.11
16. b - 240.54(C)
17. b - 430.72(B)(2) – Table 430.72(B)
18. c - 240.15(B)(1)
19. a - 240.101(A)
20. b - 700.32
21. c - 430.122(A)
22. a - 440.22(A)
23. d - 680.22(A)(4)
24. b - 240.5(B)(2)
25. b - 240.10
26. b - 240.4(F) – 450.3(B) Note 1-240.6(A)
27. d - 240.4(F)

$$\frac{secondary}{primary} = \frac{120V.}{240V.} = \frac{1}{2} = .5$$

$$\frac{300\,ampere\ primary\ overcurrent\ device}{.5} = 600\,A$$

28. c - 230.90
29. c - 408.36(D)
30. d - 408.36(A)
31. a - 240.60(D)
32. c - 240.85
33. b - 517.73(A)(1)-660.6(A)
34. b - 647.7(A)
35. b - 240.60(C)
36. c - 240.67(B)
37. d - 240.50(E)
38. a - 410.130(E)(1)
39. d - 393.45(A)
40. c - 240.6(A) –Table 240.6(A)
41. d - 551.71(C)
42. a - 550.32(C)
43. b - 455.7(A)
44. a - 702.4(A)
45. b - 680.21(C)
46. c - 760.43
47. b - 760.41(B)
48. d - 555.19(A)(4)
49. a - 690.41(B) – 690.41(B)(1)
50. d - Article 100 – Definition-Overcurrent Protective Device-Branch-Circuit

Index

About the Author

It started on August 3, 1964 as I started to work in the electrical trade through the auspices of Local 52, IBEW in Newark, N.J. Seven months later, I became a member of this Local, which was merged with Local 164 (Paramus, N.J.) in 2000.

I served my apprenticeship until March 1, 1969, at which time I became a Journeyman-Wireman. I gained a wealth of knowledge and experience during this period and the years that followed.

In March of 1978, I developed an association with a training/consulting company in Trenton, N.J. This company had a request from a client in Philadelphia to present an electrical training course for their maintenance personnel. The class was scheduled for four weeks. I was asked to conduct this course, and despite not having any teaching experience, I decided to become an instructor. Fortunately, the class went well, and the training was extended for another four weeks. In the meantime, this training/consulting company offered me a permanent position.

In September of 1978, I started to offer courses on the topic of the National Electrical Code. The key was to find a method of instruction that would benefit the students by keeping their attention during the three day course period. And, to foster an interest in this complex document that would serve them well beyond our brief time together. In later years, I developed and presented courses on the topics of Grounding Electrical Distribution Systems, Designing Overcurrent Protection, Electrical Systems In Hazardous (Classified) Locations, and Electrical Equipment Maintenance. These courses were offered by my company, Electrical Design Institute, and several universities, including the University of Wisconsin, George Washington University, North Carolina State University, the University of Toledo, and the University of Alabama.

In 2021, I authored books entitled, The NEC and You, Perfect Together, Grounding Electrical Distribution Systems, and Designing Overcurrent Protection, NEC Article 240 and Beyond. These books are published by River Publishers.

Gregory P. Bierals
May 24, 2021